PROBLEM SOLVING
in
MATHEMATICS

ISBN 0-86651-184-9

Order Number DS01410

bcdefghi-MA-898765

DALE
SEYMOUR
PUBLICATIONS
P.O. BOX 10888
PALO ALTO, CA 94303

PROBLEM SOLVING IN MATHEMATICS

PROJECT STAFF

DIRECTOR: OSCAR SCHAAF, UNIVERSITY OF OREGON
ASSOCIATE DIRECTOR: RICHARD BRANNAN, LANE EDUCATION SERVICE DISTRICT

WRITERS: RICHARD BRANNAN
                             MARYANN DEBRICK
                             JUDITH JOHNSON
                             GLENDA KIMERLING
                             SCOTT McFADDEN
                             JILL McKENNEY
                             OSCAR SCHAAF
                             MARY ANN TODD

PRODUCTION: MEREDITH SCHAAF
                             BARBARA STOEFFLER

EVALUATION: HENRY DIZNEY
                             ARTHUR MITTMAN
                             JAMES ELLIOTT
                             LESLIE MAYES
                             ALISTAIR PEACOCK

PROJECT GRADUATE   FRANK DEBRICK
   STUDENTS: MAX GILLETT
                             KEN JENSEN
                             PATTY KINCAID
                             CARTER McCONNELL
                             TOM STONE

ACKNOWLEDGMENTS:

TITLE IV-C LIAISON: Ray Talbert
                         Charles Nelson

                         <u>Monitoring</u> Team

                         Charles Barker
                         Ron Clawson
                         Jeri Dickerson
                         Anthony Fernandez
                         Richard Olson
                         Ralph Parrish
                         Fred Rugh
                         Alton Smedstad

ADVISORY COMMITTEE: Mary Grace Kantowski   University of Florida
                         John LeBlanc   Indiana University
                         Richard Lesh   Northwestern University
                         Edwin McClintock   Florida International University
                         Len Pikaart   Ohio University
                         Kenneth Vos   The College of St. Catherine

A special thanks is due to the many teachers, schools, and districts within
the state of Oregon that have participated in the development and evaluation
of the project materials. A list would be lengthy and certainly someone's
name would inadvertently be omitted. Those persons involved have the project's
heartfelt thanks for an impossible job well done.

The following projects and/or persons are thanked for their willingness to
share pupil materials, evaluation materials, and other ideas.

    Don Fineran, Mathematics Consultant, Oregon Department of Education
    Frank Lester, Indiana University
    Steve Meiring, Mathematics Consultant, Ohio Department of Education
    Harold Schoen, University of Iowa
    Iowa Problem Solving Project, Earl Ockenga, Manager
    Math Lab Curriculum for Junior High, Dan Dolan, Director
    Mathematical Problem Solving Project, John LeBlanc, Director

CONTENTS

Introduction...................i-xi

I. Getting Started 3-36

    Guess And Check 5
    Look For A Pattern 11
    Make A Systematic List 19
    Make And Use A Drawing
       Or Model 25
    Eliminate Possibilities 31

II. Drill And Practice -
    Whole Number 39-74

    Hundred Dollar Words 41
    Finding Patterns 43
    Machine Hook-Ups 45
    Blank Squares 51
    Using Parentheses 53
    Digit Draw Activities 57
    Create A Problem 61
    Shortcuts 63
    Across And Down 67
    Addition And Multiplication
       Tables 71

III. Drill And Practice -
    Fractions 77-96

    Jumping Floozie 79
    Fraction Patterns 83
    Squares And Paths 85
    Creating Fraction Problems .. 87
    Smallest Answer 91
    Closer To 95

IV. Drill And Practice -
    Decimals 99-110

    Going In Circles 101
    Going Crazy With Numbers ... 103
    A Decimal Web 105
    Multiplying For Points 107
    Hone On The Range 109

V. Percent Sense 113-159

    Sense Or Nonsense 115
    Room Designs 117
    Using A Percent Finder 119
    Transparent 100-Grids(Master). 121
    Using A 100-Grid 123
    100 Percent 125
    Percents On A Number Line ... 129
    Using Charts 135
    Per Hundred 139
    I've Got Your Number 145
    Percent Shortcuts 147
    One-Percent Method 149
    Percent Allowance 153
    Percent Applications 155
    Using Percent Sense 157

VI. Factors, Multiples, And
    Primes 163-186

    Factors 165
    Common Factors 167
    Common Multiples 173
    Prime Numbers 177
    Prime Numbers Less Than 1000 . 181
    Some Prime Investigations ... 183

VII. Measurement - Volume, Area
    Perimeter 189-210

    Stacking Blocks 191
    Cubic Inches 193
    Boxing Up Cubes 197
    Floor Tiles 199
    A Pen For Barney 201
    Greatest Volume 203
    Designing Office Spaces 205
    Skyscraper 209

VIII. Probability 213-226

      Division With Dice 215
      Rolling Some Sums 217
      Flipping Coins 219
      Quiz Or No Quiz 221
      The Coin Flip Path 225

IX. Challenges 229-264

      Money In The Bank 233
      County Fair 235
      Who's Who? 237
      Jars Of Candy 239
      Change For A Dollar 241
      Stepping High 243
      Pigs And Turkeys 245
      What's The Path? 247
      Spirolaterals - 1 249
      Spirolaterals - 2 251
      Four 4's 253
      Some Circle Puzzles 255
      Pool Patterns 257
      The Boy And The Devil 259
      Connected Stamps 261
      A Shopping Puzzle 263

INTRODUCTION

What is PSM?

PROBLEM SOLVING IN MATHEMATICS is a program of problem-solving lessons and teaching techniques for grades 4–8 and (9) algebra. Each grade-level book contains approximately 80 lessons and a teacher's commentary with teaching suggestions and answer key for each lesson. *Problem Solving in Mathematics* is not intended to be a complete mathematics program by itself. Neither is it supplementary in the sense of being extra credit or to be done on special days. Rather, it is designed to be integrated into the regular mathematics program. Many of the problem-solving activities fit into the usual topics of whole numbers, fractions, decimals, percents, or equation solving. Each book begins with lessons that teach several problem-solving skills. Drill and practice, grade-level topics, and challenge activities using these problem-solving skills complete the book.

PROBLEM SOLVING IN MATHEMATICS is designed for use with all pupils in grades 4–8 and (9) algebra. At-grade-level pupils will be able to do the activities as they are. More advanced pupils may solve the problems and then extend their learning by using new data or creating new problems of a similar nature. Low achievers, often identified as such only because they haven't reached certain computational levels, should be able to do the work in PSM with minor modifications. The teacher may wish to work with these pupils at a slower pace using more explanations and presenting the material in smaller doses.

[Additional problems appropriate for low achievers are contained in the *Alternative Problem Solving in Mathematics* book. Many of the activities in that book are similar to those in the regular books except that the math computation and length of time needed for completion are scaled down. The activities are generally appropriate for pupils in grades 4–6.]

Why Teach Problem Solving?

Problem solving is an ability people need throughout life. Pupils have many problems with varying degrees of complexity. Problems arise as they attempt to understand concepts, see relationships, acquire skills, and get along with their peers, parents, and teachers. Adults have problems, many of which are associated with making a living, coping with the energy crisis, living in a nation with peoples from different cultural backgrounds, and preserving the environment. Since problems are so central to living, educators need to be concerned about the growth their pupils make in tackling problems.

What Is a Problem?

MACHINE HOOK-UPS

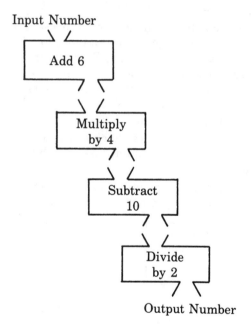

It is highly recommended that teachers intending to use *Problem Solving in Mathematics* receive training in implementing the program. The *In-Service Guide* contains much of this valuable material. In addition, in-service audio cassette tapes are available. These provide indepth guidance on using the PSM grade-level books and an overall explanation of how to implement the whole program. The tapes are available for loan upon request. Please contact Dale Seymour Publications, Box 10888, Palo Alto, CA 94303 for further information about the tapes and other possible in-service opportunities.

	Input Number	Output Number
a.	4	
b.	8	
c.	12	
d.		39
e.		47
f.		61

Suppose a 6th grader were asked to fill in the missing output blanks for a, b, and c in the table. Would this be a problem for him? Probably not, since all he would need to do is to follow the directions. Suppose a 2nd-year algebra student were asked to fill in the missing input blank for d. Would this be a problem? Probably not, since she would write the suggested equation,

$$\frac{4\,(x + 6) - 10}{2} = 39$$

and then solve it for the input. Now suppose the 6th grader were asked to fill in the input for d, would this be a problem for him? Probably it *would* be. He has no directions for getting the answer. However, if he has the desire, it is within his power to find the answer. What might he do? Here are some possibilities:

1. He might make *guesses*, do *checking*, and then make refinements until he gets the answer.

2. He might fill in the output numbers that correspond to the input numbers for a, b, and c.

	Input Number	Output Number
a.	4	
b.	8	
c.	12	
d.		39

and then observe this pattern:
For an increase of 4 for the input, the output is increased by 8.
Such an observation should lead quickly to the required input of 16.

3. He might start with the output and *work backwards* through the machine hook-up using the inverse (or opposite) operations.

For this pupil, there was no "ready-made" way for him to find the answer, but most motivated 6th-grade pupils would find a way.

A *problem*, then, is a situation in which an individual or group accepts the challenge of performing a task for which there is no immediately obvious way to determine a solution. Frequently, the problem can be approached in many ways. Occasionally, the resulting investigations are nonproductive. Sometimes they are so productive as to lead to many different solutions or suggest more problems than they solve.

What Does Problem Solving Involve?

Problem solving requires the use of many *skills*. Usually these skills need to be used in certain combinations before a problem is solved. A combination of skills used in working toward the solution of a problem can be referred to as a *strategy*. A successful strategy requires the individual or group to generate the information needed for solving the problem. A considerable amount of creativity can be involved in generating this information.

What Problem-Solving Skills Are Used in PSM?

Skills are the building blocks used in solving a problem. The pupil materials in the PSM book afford many opportunities to emphasize problem-solving skills. A listing of these skills is given below.

THE PSM CLASSIFIED LIST OF PROBLEM-SOLVING SKILLS

A. Problem Discovery, Formulation
1. State the problem in your own words.
2. Clarify the problem through careful reading and by asking questions.
3. Visualize an object from its drawing or description.
4. Follow written and/or oral directions.

B. Seeking Information
5. Collect data needed to solve the problem.
6. Share data and results with other persons.
7. Listen to persons who have relevant knowledge and experiences to share.
8. Search printed matter for needed information.
9. Make necessary measurements for obtaining a solution.
10. Record solution possibilities or attempts.
11. Recall and list related information and knowledge.

C. Analyzing Information
12. Eliminate extraneous information.
13. Find likenesses and differences and make comparisons.
14. Classify objects or concepts.
15. Make and use a drawing or model.
16. Make and/or use a systematic list or table.
17. Make and/or use a graph.
18. Look for patterns and/or properties.
19. Use mathematical symbols to describe situations.
20. Break a problem into manageable parts.

D. Solve—Putting It Together—Synthesis
21. Make predictions, conjectures, and/or generalizations based upon data.
22. Make decisions based upon data.
23. Make necessary computations needed for the solution.
24. Determine limits and/or eliminate possibilities.
25. Make reasonable estimates.
26. Guess, check, and refine.
27. Solve an easier but related problem. Study solution process for clues.
28. Change a problem into one you can solve. (Simplify the problem.)
29. Satisfy one condition at a time.
30. Look at problem situation from different points of view.
31. Reason from what you already know. (Deduce.)
32. Work backwards.
33. Check calculated answers by making approximations.
34. Detect and correct errors.
35. Make necessary measurements for checking a solution.

36. Identify problem situation in which a solution is not possible.
37. Revise the conditions of a problem so a solution is possible.

E. Looking Back—Consolidating Gains
38. Explain how you solved a problem.
39. Make explanations based upon data.
40. Solve a problem using a different method.
41. Find another answer when more than one is possible.
42. Double check solutions by using some formal reasoning method (mathematical proof).
43. Study the solution process.
44. Find or invent other problems which can be solved by certain solution procedures.
45. Generalize a problem solution so as to include other solutions.

F. Looking Ahead—Formulating New Problems
46. Create new problems by varying a given one.

What Are Some Examples of Problem-Solving Strategies?

Since strategies are a combination of skills, a listing (if it were possible) would be even more cumbersome than the list of skills. Examples of some strategies that might be used in the "Machine Hook-Ups" problem follow:

Strategy 1. *Guess* the input; *check* by computing the output number for your guess; if guess does not give the desired output, note the direction of error; *refine* the guess; compute; continue making refinements until the correct output results.

Strategy 2. *Observe* the *patterns* suggested by the input and output numbers for the *a, b, c* entries in the table; *predict* additional output and input numbers by extending both patterns; *check* the predicted input for the *d* entry by computing.

Strategy 3. *Study* the operations suggested in the machine hook-up; *work backwards* through the machine *using previous knowledge* about inverse operations.

An awareness of the strategies being used to solve a problem is probably the most important step in the development of a pupil's problem-solving abilities.

What is the Instructional Approach Used in PSM?

The content objectives of the lessons are similar to those of most textbooks. The difference is in the approach used. First, a wider variety of problem-solving skills is emphasized in the materials than in most texts. Second, different styles of teaching such as direct instruction, guided discovery, laboratory work, small-group discussions, nondirective instruction, and individual work all have a role to play in problem-solving instruction.

Most texts employ direct instruction almost exclusively, whereas similar lessons in PSM are patterned after a guided discovery approach. Also, an attempt is made in the materials to use intuitive approaches extensively before teaching formal algorithms. Each of the following is an integral part of the instructional approach to problem solving.

A. TEACH PROBLEM-SOLVING SKILLS DIRECTLY

Problem-solving skills such as "follow directions," "listen," and "correct errors" are skills teachers expect pupils to master. Yet, such skills as "guess and check," "make a systematic list," "look for a pattern," or "change a problem into one you can solve" are seldom made the object of direct instruction. These skills, as well as many more, need emphasis. Detailed examples for teaching these skills early in the school year are given in the commentaries to the *Getting Started* activities.

B. INCORPORATE A PROBLEM-SOLVING APPROACH WHEN TEACHING TOPICS IN THE COURSE OF STUDY

Drill and practice activities. Each PSM book includes many pages of drill and practice at the problem-solving level. These pages, along with the *Getting Started* section, are easy for pupils and teachers to get into and should be started early in the school year.

Laboratory activities and investigations involving mathematical applications and readiness activities. Readiness activities from such mathematical strands as geometry, number theory, and probability are included in each book. For example, area explorations are used in grade 4 as the initial stage in the teaching of the multiplication and division algorithms and fraction concepts.

Teaching mathematical concepts, generalizations, and processes. Each book includes two or more sections on grade-level content topics. For the most part, these topics are developmental in nature and usually need to be supplemented with practice pages selected from a textbook.

C. PROVIDE MANY OPPORTUNITIES FOR PUPILS TO USE THEIR OWN PROBLEM-SOLVING STRATEGIES

One section of each book includes a collection of challenge activities which provide opportunities for emphasizing problem-solving strategies. Generally, instruction should be nondirective, but at times suggestions may need to be given. If possible, these suggestions should be made in the form of alternatives to be explored rather than hints to be followed.

D. CREATE A CLASSROOM ATMOSPHERE IN WHICH OPENNESS AND CREATIVITY CAN OCCUR

Such a classroom climate should develop if the considerations mentioned in A, B, and C are followed. Some specific suggestions to keep in mind as the materials are used are:

- Set an example by solving problems and by sharing these experiences with the pupils.
- Reduce anxiety by encouraging communication and cooperation. On frequent occasions problems might be investigated using a cooperative mode of instruction along with brainstorming sessions.
- Encourage pupils in their efforts to solve a problem by indicating that their strategies are worth trying and by providing them with sufficient time to investigate the problem; stress the value of the procedures pupils use.
- Use pupils' ideas (including their mistakes) in solving problems and developing lessons.
- Ask probing questions which make use of words and phrases such as
  I wonder if . . . .
  Do you suppose that . . . .
  What happens if . . . .
  How could we find out . . . .
  Is it possible that . . . .
- Reinforce the asking of probing questions by pupils as they search for increased understanding. Pupils seldom are skilled at seeking probing questions but they can be taught to do so. If instruction is successful, questions of the type, "What should I do now?," will be addressed to themselves rather than to the teacher.

What Are the Parts of Each PSM Book?

PROBLEM SOLVING IN MATHEMATICS

Grade 4	Grade 5	Grade 6	Grade 7	Grade 8	Grade 9
Getting Started	Getting Started	Getting Started	Getting Started	Getting Started	Getting Started
Place Value Drill and Practice	Whole Number Drill and Practice	Drill and Practice	Drill and Practice– Whole Numbers	Drill and Practice	Algebraic Concepts and Patterns
Whole Number Drill and Practice	Story Problems	Story Problems	Drill and Practice– Fractions	Variation	Algebraic Explanations
Multiplication and Division Concepts	Fractions	Fractions	Drill and Practice– Decimals	Integer Sense	Equation Solving
Fraction Concepts	Geometry	Geometry	Percent Sense	Equation Solving	Word Problems
Two-digit Multiplication	Decimals	Decimals	Factors, Multiples, and Primes	Protractor Experiments	Binomials
Geometry	Probability	Probability	Measurement–Volume, Area, Perimeter	Investigations in Geometry	Graphs and Equations
Rectangles and Division	Estimation with Calculators	Challenges	Probability	Calculator	Graph Investigations
Challenges	Challenges		Challenges	Percent Estimation	Systems of Linear Equations
				Probability	Challenges
				Challenges	

Notice that the above chart is only a scope of PSM—not a scope and sequence. In general, no sequence of topics is suggested with the exceptions that *Getting Started* activities must come early in the school year and *Challenge* activities are usually deferred until later in the year.

Getting Started Several problem-solving skills are presented in the *Getting Started* section of each grade level. Hopefully, by concentrating on these skills during the first few weeks of school pupils will have confidence in applying them to problems that occur later on. In presenting these skills, a direct mode of instruction is recommended. Since the emphasis needs to be on the problem-solving skill used to find the solution, about ten to twelve minutes per day are needed to present a problem.

Drill And Practice No sequence is implied by the order of activities included in these sections. They can be used throughout the year but are especially appropriate near the beginning of the year when the initial chapters in the textbook emphasize review. Most of the activities are not intended to develop any particular concept. Rather, they are drill and practice lessons with a problem-solving flavor.

Challenges Fifteen or more challenge problems are included in each book. In general, these should be used only after *Getting Started* activities have been completed and pupils have had some successful problem-solving experiences.

Many of the other sections in PSM are intended to focus on particular grade-level content. The purpose is to provide intuitive background for certain topics. A more extensive textbook treatment usually will need to follow the intuitive development.

Teacher Commentaries Each section of a PSM book has an overview teacher commentary. The overview commentary usually includes some philosophy and some suggestions as to how the activities within the section should be used. Also, every pupil page in PSM has a teacher commentary on the back of the lesson. Included here are mathematics teaching objectives, problem-solving skills pupils might use, materials needed, comments and suggestions, and answers.

How Often Should Instruction Be Focused on Problem Solving?

Some class time should be given to problem solving nearly every day. On some days an entire class period might be spent on problem-solving activities; on others, only 8 to 10 minutes. Not all the activities need to be selected from PSM. Your textbook may contain ideas. Certainly you can create some of your own. Many companies now have published excellent materials which can be used as sources for problem-solving ideas. Frequently, short periods of time should be used for identifying and comparing problem-solving skills and strategies used in solving problems.

How Can I Use These Materials When I Can't Even Finish What's in the Regular Textbook?

This is a common concern. But PSM is not intended to be an "add-on" program. Instead, much of PSM can replace material in the textbook. Correlation charts can be made suggesting how PSM can be integrated into the course of study or with the adopted text. Also, certain textbook companies have correlated their tests with the PSM materials.

Can the Materials Be Duplicated?

The pupil lessons may be copied for students. Each pupil lesson may be used as an overhead projector transparency master or as a blackline duplicator master. Sometimes the teacher may want to project one problem at a time for pupils to focus their attentions on. Other times, the teacher might want to duplicate a lesson for individual or small group work. Permission to duplicate pupil lesson pages for classroom use is given by the publisher.

How Can a Teacher Tell Whether Pupils Are Developing and Extending Their Problem-Solving Abilities?

Presently, reliable paper and pencil tests for measuring problem-solving abilities are not available. Teachers, however, can detect problem-solving growth by observing such pupil behaviors as

- identifying the problem-solving skills being used.

- giving accounts of successful strategies used in working on problems.
- insisting on understanding the topics being studied.
- persisting while solving difficult problems.
- working with others to solve problems.
- bringing in problems for class members and teachers to solve.
- inventing new problems by changing problems previously solved.

What Evidence Is There of the Effectiveness of PSM?

Although no carefully controlled longitudinal study has been made, evaluation studies do indicate that pupils, teachers, and parents like the materials. Scores on standardized mathematics achievement tests show that pupils are registering greater gains than expected on all parts of the test, including computation. Significant gains were made on special problem-solving skills tests which were given at the beginning and end of a school year.

Also, when selected materials were used exclusively over a period of several weeks with 6th-grade classes, significant gains were made on the word-problem portion of the standardized test. In general, the greater gains occurred in those classrooms where the materials were used as specified in the teacher commentaries and in-service materials.

Teachers have indicated that problem-solving skills such as *look for a pattern, eliminate possibilities*, and *guess and check* do carry over to other subjects such as Social Studies, Language Arts, and Science. Also, the materials seem to be working with many pupils who have not been especially successful in mathematics. And finally, many teachers report that PSM has caused them to make changes in their teaching style.

Why Is It Best to Have Whole-Staff Commitment?

Improving pupils' abilities to solve problems is not a short-range goal. In general, efforts must be made over a long period of time if permanent changes are to result. Ideally, then, the teaching staff for at least three successive grade levels should commit themselves to using PSM with their pupils. Also, if others are involved, this will allow for opportunities to plan together and to share experiences.

How Much In-Service Is Needed?

A teacher who understands the meaning of problem solving and is comfortable with the different styles of teaching it requires could get by with self in-service by carefully studying the section and page commentaries in a grade-level book. The different styles of teaching required include direct instruction, guided discovery, laboratory work, small group instruction, individual work, and nondirective instructions. The teacher would find the audio tapes for each book and the *In-Service Guide* a valuable resource and even a time saver.

If a school staff decides to emphasize problem solving in all grade levels where PSM books are available, in-service sessions should be led by someone who has used the materials in the intended way. For more information on this in-service see the *In-Service Guide*.

PROBLEM-SOLVING PROGRAM

IV-C

REQUIRED MATERIALS	Grade 4	5	6	7	8	9
blank cards	X	X	X	X	X	X
bottle caps or markers	X			X		
calendar						X
calculators (optional for some activities)	X	X	X	X	X	X
cm squared paper, strips and singles						X
coins				X	X	
colored construction paper (circle fractions)	X	X				
cubes		X	X	X		X
cubes with red, yellow and green faces					X	
Cuisenaire rods (orange and white) or strips of paper		X				
dice (blank wooden or foam, for special dice)	X					
dice, regular (average 2 per student)	X	X	X	X	X	
geoboards, rubber bands, and record paper	X		X			
graph paper or cm squared paper				X		X
grid paper (1")			X			
metric rulers				X	X	X
phone books, newspapers, magazines		X				
protractors and compasses					X	
scissors	X	X	X	X		
spinners (2 teacher-made)			X			
tangrams	X					
tape measures				X		
thumbtacks (10 per pair of students)		X				
tile	X		X			
tongue depressors	X					
uncooked spaghetti or paper strips			X			

PSM   Rev. 1982

RECOMMENDED MATERIALS	Grade 4	5	6	7	8	9
adding machine tape				X		
centimetre rulers			X	X		
colored pens, pencils, or crayons		X				
coins, toy or real	X					
coins (two and one-half)						X
cubes					X	
demonstration ruler for overhead		X				
dominoes					X	
geoboard, transparent (for overhead)	X		X			
money - 20 $1.00 bills per student			X			
moveable markers		X	X		X	
octahedral die for extension activity			X	X		
overhead projector	X	X	X	X	X	X
place value frame and markers			X			
straws, uncooked spaghetti, or toothpicks		X				
transparent circle fractions for overhead	X	X				

Where Can I Find Other Problem Solving Materials?

RESOURCE BIBLIOGRAPHY

The number in parentheses refers to the list of publishers on the next page.

For <u>students</u> <u>and</u> <u>teachers</u>:

AFTER MATH, BOOKS I—IV by Dale Seymour, et al.
    Puzzles to solve -- some of them non-mathematical.   (1)

AHA, INSIGHT by Martin Gardner
    Puzzles to solve -- many of them non-mathematical.   (3)

THE BOOK OF THINK by Marilyn Burns
    Situations leading to a problem-solving investigation.   (1)

CALCULATOR ACTIVITIES FOR THE CLASSROOM, BOOKS 1 & 2 by George Immerzeel and
    Earl Ockenga
    Calculator activities using problem solving.   (1)

GEOMETRY AND VISUALIZATION by Mathematics Resource Project
    Resource materials for geometry.   (1)

GOOD TIMES MATH EVENT BOOK by Marilyn Burns
    Situations leading to a problem-solving investigation.   (1)

FAVORITE PROBLEMS by Dale Seymour
    Problem solving challenges for grades 5-7.   (3)

FUNTASTIC CALCULATOR MATH by Edward Beardslee
    Calculator activities using problem solving.   (4)

I HATE MATHEMATICS! BOOK by Marilyn Burns
    Situations leading to a problem solving investigation.   (3)

MATHEMATICS IN SCIENCE AND SOCIETY by Mathematics Resource Project
    Resource activities in the fields of astronomy, biology, environment,
    music, physics, and sports.   (1)

MIND BENDERS by Anita Harnadek
    Logic problems to develop deductive thinking skills. Books A-1, A-2, A-3,
    and A-4 are easy. Books B-1, B-2, B-3, and B-4 are of medium difficulty.
    Books C-1, C-2, and C-3 are difficult.   (6)

NUMBER NUTZ (Books A, B, C, D) by Arthur Wiebe
    Drill and practice activities at the problem solving level.   (2)

NUMBER SENSE AND ARITHMETIC SKILLS by Mathematics Resource Project
    Resource materials for place value, whole numbers, fractions, and decimals.   (1)

The <u>Oregon</u> <u>Mathematics</u> <u>Teacher</u> (magazine)
    Situations leading to a problem solving investigation.   (8)

PROBLEM OF THE WEEK by Lyle Fisher and William Medigovich
    Problem solving challenges for grades 7-12.   (3)

RATIO, PROPORTION AND SCALING by Mathematics Resource Project
    Resource materials for ratio, proportion, percent, and scale drawings.   (1)

STATISTICS AND INFORMATION ORGANIZATION by Mathematics Resource Project
    Resource materials for statistics and probability.   (1)

SUPER PROBLEMS by Lyle Fisher
    Problem solving challenges for grades 7-9.   (3)

For teachers only:

DIDACTICS AND MATHEMATICS by Mathematics Resource Project   (1)

HOW TO SOLVE IT by George Polya   (3)

MATH IN OREGON SCHOOLS by the Oregon Department of Education   (9)

PROBLEM SOLVING: A BASIC MATHEMATICS GOAL by the Ohio Department of Education   (3)

PROBLEM SOLVING: A HANDBOOK FOR TEACHERS by Stephen Krulik and Jesse Rudnik   (1)

PROBLEM SOLVING IN SCHOOL MATHEMATICS by NCTM   (7)

Publisher's List

1. Creative Publications, 3977 E Bayshore Rd, PO Box 10328, Palo Alto, CA 94303

2. Creative Teaching Associates, PO Box 7714, Fresno, CA 93727

3. Dale Seymour Publications, PO Box 10888, Palo Alto, CA 94303

4. Enrich, Inc., 760 Kifer Rd, Sunnyvale, CA 94086

5. W. H. Freeman and Co., 660 Market St, San Francisco, CA 94104

6. Midwest Publications, PO Box 448, Pacific Grove, CA 93950

7. National Council of Teachers of Mathematics, 1906 Association Dr, Reston, VA
    22091

8. Oregon Council of Teachers of Mathematics, Clackamas High School,
    13801 SE Webster St, Milwaukie, OR 97222

9. Oregon Department of Education, 700 Pringle Parkway SE, Salem, OR 97310

Grade 7

I. GETTING STARTED

I. GETTING STARTED

Teachers usually are successful at
teaching skills in mathematics. Besides
computation skills, they emphasize skills
in following directions, listening, de-
tecting errors, explaining, recording,
comparing, measuring, sharing, ...
They (You!) can also teach problem-
solving skills. This section is designed
to help teachers teach and students learn
specific problem-solving skills.

Some Problem-Solving Skills

Five common but powerful problem-solving skills are introduced in this
section. They are: . guess and check
                          . look for a pattern
                          . make a systematic list
                          . make and use a drawing or model
                          . eliminate possibilities

Students might use other skills to solve the problems. They can be
praised for their insight but it is usually a good idea to limit the list
of skills taught during the first few lessons. More problem-solving skills
will occur in the other sections.

An Important DON'T

When you read the episodes that follow in this Getting Started section
notice how the lessons are very teacher directed. The main purpose is to
teach the problem-solving skills. Teachers should stress the skills
verbally and write them on the board. Don't just ditto these activities
and hand them out to be worked. Teacher direction through questions,
summaries, praise, etc. is most important for teaching the problem-solving
skills in this section. We want students to focus on specific skills
which will be used often in all the sections. Later, in the Challenge
Problems section, students will be working more independently.

Using The Activities

If you heed the important Don't on the previous page, you are on your way to success! The problems here should fit right in with your required course of study as they use whole number skills, elementary geometry and money concepts. In most cases, students will have the prerequisites for the problems in this section although you might want to check over each problem to be sure.

No special materials are required although markers, coins and dice are helpful for some of the problems. The large type used for the problems makes them easier to read if they are shown on an overhead screen. In most cases students can easily copy the problem from the overhead. At other times you might copy the problem onto the chalkboard.

When And How Many

The Getting Started section should be used at the beginning of the year as it builds background in problem-solving skills for the other sections. As the format indicates only one problem per day should be used. Each should take less than twelve minutes of classtime if the direct mode of instruction is used. The remainder of the period is used for a lesson from the textbook or perhaps an activity from the Drill and Practice section of these materials.

REMEMBER: One Problem Per Day when you are using this Getting Started section.

Guess And Check

The episode that follows shows how one teacher teaches the skill of guess and check. Notice that she very closely directs the instruction and constantly uses the terminology.

Ms. Adams: Who am I? If you multiply me by 5 and then subtract 8, the result is 52. I wonder if the number is 10. Is it? (Waits for hands.) Jeff?

Jeff: Nope!

Ms. A: How do you know?

Jeff: 'Cause 10 times 5 is 50, subtract 8 gives 42.

Ms. A: By checking my guess you found out it was off. Well, is the number 15? Julie?

Julie: (Thinking out loud.) 15 times 5 is 75, subtract 8 is 67. No - too big.

Ms. A: Fifteen is too big? What can you say about 10?

Jeff: It was too small.

Ms. A: Guessing and checking helped you decide 10 is too small and and 15 is too big. Let's refine the guesses. Refine means to make better guesses. Can you make a better guess? (Some puzzled, some thinking, some hands.) Larry?

Larry: The answer should probably be closer to 10 than 15. Try 12.

Joyce: Yes, that works. Twelve times 5 is 60, subtract 8 is 52.

Ms. A: Guess, check and refine is a good way to solve problems. We're going to use it a lot this year. I'm putting it up on the wall so we'll all remember how important it is! Let's try another problem.....

GUESS AND CHECK

WEEK 1 - DAY 1

a. Who am I? If you multiply me by 5 and then subtract 8, the result is 52.

b. Who am I? If you multiply me by 15 and add 28, the result is 103.

c. Who am I? If you divide me by 4 and subtract 16, the result is 4.

****

WEEK 1 - DAY 2

Suzie has 2 more dimes than nickels. This amounts to $1.85. How many of each coin does she have?

****

WEEK 1 - DAY 3

Study this pattern: 3, 4, 7, 11, 18, 29, 47, 76

Note that 3 + 4 = 7 and 4 + 7 = 11 and 7 + 11 = 18, etc.

Use the same rule to complete the patterns below.

a. 5, 11, 16, ___, ___, ___, ___

b. 2, ___, 8, ___, ___, ___, ___,

c. 3, ___, ___, 13, ___, ___, ___

d. 1, ___, ___, ___, ___, 38

<u>Guess</u> <u>And</u> <u>Check</u>

Day 1. Answers: a. 12   b. 5   c. 80

Comments and suggestions:

. This problem gives you opportunity to demonstrate the "refinement"
  process. For example, in part (a), first show that "10" is too
  small and "15" is too large. The answer must be between 10 and
  15.

Day 2. Answer: 11 nickels and 13 dimes

Comments and suggestions:

. Pupils can get many combinations of nickels and dimes that total
  $1.85. They need to be reminded of the "2 more dimes than nic-
  kels" condition. Once again the teacher has the opportunity to
  stress the refinement process. For example, is the combination
  of 9 dimes and 7 nickels too large or too small? How could we
  refine this guess?

Day 3. Answers: a. 5, 11, 16, <u>27</u>, <u>43</u>, <u>70</u>, <u>113</u>

                    b. 2, <u>6</u>, 8, <u>14</u>, <u>22</u>, <u>36</u>, <u>58</u>

                    c. 3, <u>5</u>, <u>8</u>, 13, <u>21</u>, <u>34</u>, <u>55</u>

                    d. 1, <u>7</u>, <u>8</u>, <u>15</u>, <u>23</u>, 38

Comments and suggestions:

. Emphasize the guess, check, and refine process necessary to
  obtain solutions to parts (c) and (d). After pupils have
  solved these four problems, you might have them create some
  of their own and have their classmates solve them.

Guess And Check (cont.)

WEEK 1 - DAY 4

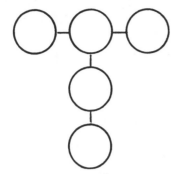

Use the numbers from 1-5.
Use them only once.
The sum must be the same in both directions.

Try to find more than one solution.

****

WEEK 1 - DAY 5

Study the example. Note that

$2 + 8 = 10$
$2 + 5 = 7$
$5 + 3 = 8$
$3 + 8 = 11$

Example

$$\boxed{2} \quad 10 \quad \boxed{8}$$
$$7 \qquad\qquad 11$$
$$\boxed{5} \quad 8 \quad \boxed{3}$$

Use the same procedure. If possible, find numbers for the circles in each problem.

a.

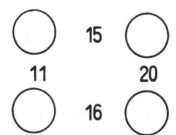

15

11   20

16

b.

12

20   11

19

c.

9

12   4

8

PSM 81

-9-

Guess And Check

Day 4. Answers: Three solutions are possible.

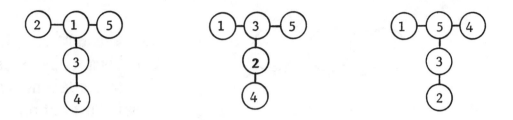

Comments and suggestions:

. You might suggest an initial guess of 2 for the "middle number." Pupils will soon realize that this guess will not lead to a solution and will be forced into trying other guesses for the middle number.

. Notice that the "magic sums" for this problem are 8, 9, and 10. What if we extended the problem and used the numbers 2 to 6 rather than 1 to 5? Now what are the possible solutions and magic sums? Some interesting patterns show up.

Day 5. Answers: a. & b. An unlimited number of solutions.
c. No answer.

Comments and suggestions:

. Most pupils will be delighted to find that their initial guesses work for (a) and (b). In fact, it may be interesting for them to try to find an initial guess that does not work.

. Part (c) does not have a solution. Pupils will attempt to make reasonable refinements but with no success. After they have made several attempts, see if they can change one number in the problem so that it will work out.

. Some pupils may be able to discover why the first two have solutions. Perhaps they can then create other similar problems.

<u>Look</u> <u>For</u> <u>A</u> <u>Pattern</u>

By now your pupils are familiar with the skill <u>guess</u> and <u>check</u>.
One teacher introduced the next problem-solving skill <u>look</u> <u>for</u> <u>a</u>
<u>pattern</u> in this way.

Mr. Jones: Who remembers what method we used to solve problems last week?

Donna: We guessed.

Mr. Jones: Is that all?

Bob: Guess and check! It's up on the poster!

Mr. Jones: That's right and we're going to add another problem-solving
            skill to the poster today. (Writes it up.)   What does it say?

Class: Look for a pattern.

Mr. Jones: This week we are going to practice looking for patterns. Here's
            our problem. (Shows problem on overhead.)   What goes in the
            next three spaces of the table?

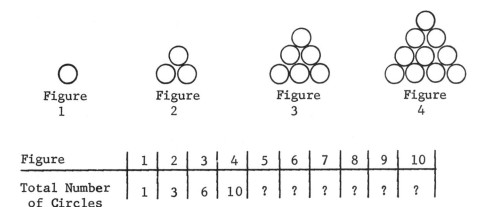

Figure	Figure 1	Figure 2	Figure 3	Figure 4

Figure	1	2	3	4	5	6	7	8	9	10
Total Number of Circles	1	3	6	10	?	?	?	?	?	?

Bob: 15, 21, 28

Mr. Jones: (Writes the numbers in the proper spaces.)   How did you arrive
            at these results, Bob?

Bob: I noticed a pattern in the bottom row. The numbers jump by 2,
            then by 3, then by 4, etc.

Mr. Jones: What are the next three numbers?   Tammi?

Tammi: 36, 45, 55. But I used a different pattern than Bob. To get
            the number of circles you can add the two numbers in the previous
            column plus 1.

Mr. Jones: In other words, to get the number in the 5th column you'd add
            4 + 10 + 1. That's an excellent procedure. Are there others?
            (Discuss other patterns that pupils observe.)

LOOK FOR A PATTERN

Study the figures below. Then complete the chart.
Look for patterns.

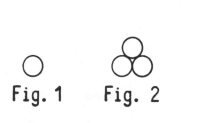 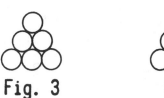

Fig. 1   Fig. 2   Fig. 3   Fig. 4

Figure	1	2	3	4	5	6	7	8	9	10
Total Number of Circles	1	3	6	10	?	?	?	?	?	?

****

WEEK 2 - DAY 2

$$1 \times 9 + 2 = \underline{\hspace{1cm}}$$

$$12 \times 9 + 3 = \underline{\hspace{1cm}}$$

$$123 \times 9 + 4 = \underline{\hspace{1cm}}$$

$$\underline{\hspace{1cm}} \times \underline{\hspace{1cm}} + \underline{\hspace{1cm}} = \underline{\hspace{1cm}}$$

$$\underline{\hspace{1cm}} \times \underline{\hspace{1cm}} + \underline{\hspace{1cm}} = \underline{\hspace{1cm}}$$

Complete the pattern.

PSM 81

Day 1. Answers:

Figure	1	2	3	4	5	6	7	8	9	10
Total Number of Circles	1	3	6	10	15	21	28	36	45	55

Comments and suggestions:

. An outline of how to introduce this problem is given on page 11.
  Remember, these problems are meant to be teacher directed.

. The numbers that appear in the bottom row of the table are
  called triangular numbers and occur frequently in mathematics.

Day 2. Answers:

$$1 \times 9 + 2 = \underline{11}$$
$$12 \times 9 + 3 = \underline{111}$$
$$123 \times 9 + 4 = \underline{1,111}$$
$$\underline{1,234 \times 9 + 5} = \underline{11,111}$$
$$\underline{12,345 \times 9 + 6} = \underline{111,111}$$

Comments and suggestions:

. Some pupils may be curious as to how far this pattern can be
  extended. They should be encouraged to make conjectures and
  then actually check them by computation.

Look For A Pattern (cont.)

WEEK 2 - DAY 3

Complete these patterns:

a. 2, 5, 8, 11, ___, ___, ___, ___

b. 1, 6, 11, 16, ___, ___, ___, ___

c. 64, 32, 16, 8, ___, ___, ___, ___

d. 1, 2, 4, 7, ___, ___, ___, ___

e. 1, 4, 9, 16, ___, ___, ___, ___

****

WEEK 2 - DAY 4

The 1st figure contains 1 square.
The 2nd figure contains 5 squares.
The 3rd figure contains 9 squares.

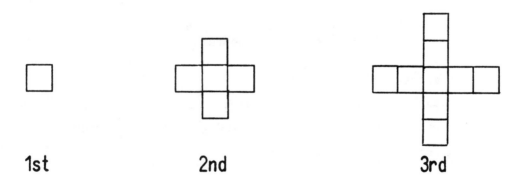

1st   2nd   3rd

If you made a drawing of the 4th figure, how many squares would it contain? How many squares would the 10th figure contain?

<u>Look</u> <u>For</u> <u>A</u> <u>Pattern</u>

Day 3. Answers: a. 2, 5, 8, 11, <u>14</u>, <u>17</u>, <u>20</u>, <u>23</u>

b. 1, 6, 11, 16, <u>21</u>, <u>26</u>, <u>31</u>, <u>36</u>

c. 64, 32, 16, 8, <u>4</u>, <u>2</u>, <u>1</u>, $\frac{1}{2}$

d. 1, 2, 4, 7, <u>11</u>, <u>16</u>, <u>22</u>, <u>29</u>

e. 1, 4, 9, 16, <u>25</u>, <u>36</u>, <u>49</u>, <u>64</u>

Comments and suggestions:

. Pupils usually enjoy exercises like these. You might make up
others for them to do or have the pupils create some of their own.

. Occasionally pupils will recognize patterns different from those
given above. These definitely should be discussed.

Day 4. Answers:

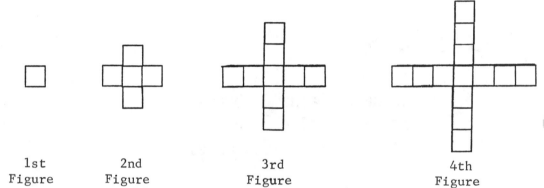

1st   2nd   3rd   4th
Figure   Figure   Figure   Figure

The 4th figure contains 13 squares.
The 10th figure would contain 37 squares.

Comments and suggestions:

. Each figure has 4 more squares than the previous one. By
extending this pattern we find that the 10th figure will
contain 37 squares.

-16-

Look For A Pattern (cont.)

WEEK 2 - DAY 5

Here's the start of a 100 chart.

1	2	3	4	5	6	7	8	9	10
11	12	13	14	15	16	17	18	19	20
21	22	23	24	25	26	27	28	29	30
					~~36~~	~~37~~	38	39	40

Without extending the chart, determine the numbers which should go in the shaded squares.

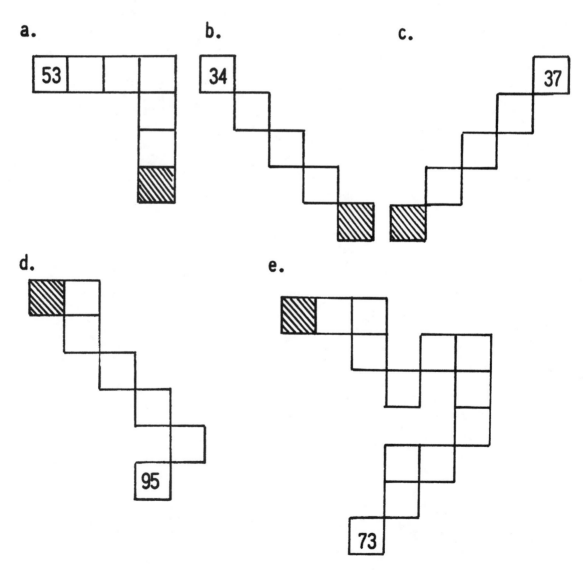

a.

b.

c.

d.

e.

Day 5. Answers: a. 86   b. 78   c. 73   d. 42   e. 11

        Comments and suggestions:

. An overhead transparency would be helpful when discussing
   this problem.

. The various patterns the pupils used to determine the answers
   should be discussed. Once again, pupils might enjoy creating
   similar problems for their classmates to solve.

Make A Systematic List

The first problem on the next page can be used to motivate the next problem-solving skill, make a systematic list. Mrs. Wilson showed the problem on an overhead.

Mrs. W: What do you notice about the organization of the list so far? Pam?

Pam: You began by listing three 10's. Then, you listed the different ways of getting two 10's.

Mrs. W: Are there any other ways of getting two 10's?

Bill: I don't think so.

Mrs. W: According to the organization in the list so far, what combinations should we work on next?

Bill: Probably those which have only one 10. (Bill adds three more to the list - see below.)

10	5	1	Total
✓✓✓			30
✓✓	✓		25
✓✓		✓	21
✓	✓✓		20
✓	✓	✓	16
✓		✓✓	12

Mrs. W: It doesn't appear as though there are any more with just one 10. Notice that in our systematic list we started with three 10's, then combinations with two 10's, and finally combinations using just one 10. What would you suggest next?

Dave: Find all combinations using no 10's and at least one 5. I think there are three of them.

Mrs. W: (Completes the organized list showing all 10 possibilities.) Do you see how making a systematic list can help us solve a problem?

MAKE A SYSTEMATIC LIST

WEEK 3 - DAY 1

Suppose you throw three darts. All of them hit the target. There are ten different totals possible. Use a systematic list, like the one below, to help you find all ten.

10	5	1	Total
✓✓✓			30
✓✓	✓		25
✓✓		✓	21

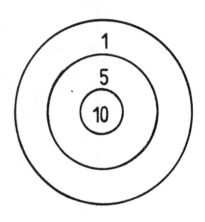

****

WEEK 3 - DAY 2

Doug was given this puzzle: The sum of two numbers is 108. One number is twice as large as the other. What are the two numbers?

Study this table. Try some other numbers.

1st number	2nd number	Sum	Is the sum 108 ?
10	20	30	No
25	50	75	No
40	80	120	No

Make A Systematic List

Day 1. Answers:

10	5	1	Total
✓✓✓			30
✓✓	✓		25
✓✓		✓	21
✓	✓✓		20
✓	✓	✓	16
✓		✓✓	12
	✓✓✓		15
	✓✓	✓	11
	✓	✓✓	7
		✓✓✓	3

Comments and suggestions:

. An outline of how to introduce this problem is given
on page 19. Remember these
problems are meant to be
teacher directed.

. Pupils need to become aware
of the value of listing information in an organized
fashion since such a list
will assist them in getting
all possibilities.

Day 2. Answers: 36 and 72

Comments and suggestions:

. Pupils need to become aware that by organizing guesses in a
table it is often easier to see the adjustments needed to make
the next guess a better one. Making tables is a tedious task
and until pupils see the value of such an endeavor, they will
not want to take the time.

Make A Systematic List (cont.)

WEEK 3 - DAY 3

Slim has $4.00 to spend on hamburgers and colas. Hamburgers cost 80¢ each. Colas cost 40¢ each. List all possible ways Slim could spend his money on hamburgers and colas.

****

WEEK 3 - DAY 4

Adam, Bill, Carl, and Dean were buying tickets to a movie. In how many different ways could they line up?

Complete this listing: A B C D
                                A B D C
                                A C B D
                                A C D B
                                A D B C
                                   ⋮

****

WEEK 3 - DAY 5

Two dice are thrown. One is red; the other is green.
List all possible ways to get each sum.

SUM	
2	1 + 1
3	1 + 2, 2 + 1
4	1 + 3, 2 + 2, 3 + 1
5	
6	
7	
8	
9	
10	
11	
12	

Day 3. Answers:

Hamburgers	5	4	3	2	1	0
Colas	0	2	4	6	8	10

Comments and suggestions:

. Pupils will no doubt list the results in different ways.
You should point out that if the listing is organized, as in
the solution above, pupils will be able to more easily detect
a pattern and therefore make it easier to complete the table.

Day 4. Answers: There are 24 possible ways:

ABCD	BACD	CABD	DABC
ABDC	BADC	CADB	DACB
ACBD	BCAD	CBAD	DBAC
ACDB	BCDA	CBDA	DBCA
ADBC	BDAC	CDAB	DCAB
ADCB	BDCA	CDBA	DCBA

Comments and suggestions:

. This might be a good time to show the beginning of a very

disorganized list: ABCD
                      BADC   . Hopefully, the class
                      BDCA
                      DABC   would soon recognize
                      CBDA
                      DCAB   the importance of
                        .
                        .   organization.
                        .

Day 5. Answers: Sum

Sum	
2	1 + 1
3	1 + 2, 2 + 1
4	1 + 3, 2 + 2, 3 + 1
5	1 + 4, 2 + 3, 3 + 2, 4 + 1
6	1 + 5, 2 + 4, 3 + 3, 4 + 2, 5 + 1
7	1 + 6, 2 + 5, 3 + 4, 4 + 3, 5 + 2, 6 + 1
8	2 + 6, 3 + 5, 4 + 4, 5 + 3, 6 + 2
9	3 + 6, 4 + 5, 5 + 4, 6 + 3
10	4 + 6, 5 + 5, 6 + 4
11	5 + 6, 6 + 5
12	6 + 6

Comments and suggestions:

. An overhead transparency will be helpful when discussing this problem.

. An important feature to discuss is the symmetrical patterns which
are apparent in the listings for the individual sums and also in the
entire display.

. Pupils will have to be reminded that an ordinary die does not have
numbers greater than 6.

Usually the problem-solving skills in this section should be introduced directly, but sometimes they can be brought out by taking advantage of pupil efforts. This problem was given in the fourth week of problem solving in Mr. Smith's class.

A ball rebounds $\frac{1}{2}$ of the height from which it is dropped. The ball is dropped from a height of 128 feet and keeps on bouncing. How far will it have traveled when it strikes the ground for the fifth time?

Pupils were allowed to think about the problem for a few minutes. Soon some thought they had solved it, others weren't sure and some had given up.

Mr. S: Steve, you wrote down some things. Will you share what you did?

Steve: Well, I first added 128, 64, 32, 16, and 8 and the result was 248. But my diagram shows that it has to be more than that.

Mr. S: I saw several others also making drawings. Do you know what our next problem-solving skill is going to be? Make and use a drawing or model. That sounds long and complicated but some of you are already doing it. Let's look at some of the drawings you tried. Sally, will you describe what you did.

Sally: I made a drawing which enabled me to keep track of each rebound.

Mr. S: Will you show us?

Sally: Yes. I thought the answer would be 248, same as Steve's. But I forgot to count both the ups and downs as you can see from my drawing.

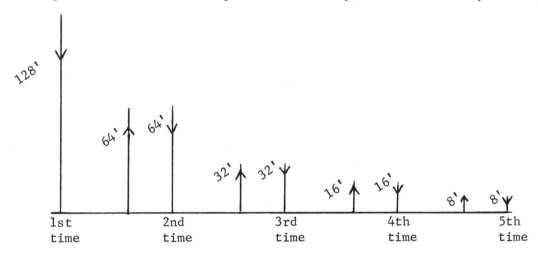

Mr. S: We see that the diagram is very helpful. What is the answer to the problem?

Class: 368 feet.

MAKE AND USE A DRAWING OR MODEL

WEEK 4 - DAY 1

A ball rebounds $\frac{1}{2}$ of the height from which it is dropped.
The ball is dropped from a height of 128 feet and keeps on
bouncing. How far will it have traveled when it strikes the
ground for the fifth time?

****

WEEK 4 - DAY 2

This shows 3 views of
a special dice.

What number is opposite the 5?
What number is opposite the 1?
What number is opposite the 2?

1st view   2nd view   3rd view

****

WEEK 4 - DAY 3

Some girls are standing in a circle. They are evenly spaced and
numbered in order. The 7th girl is directly opposite the 17th.
How many girls are in the circle?

Make And Use A Drawing Or Model

Day 1. Answer: 368 feet

Comments and suggestions:

. An outline of how this problem might be introduced is given on page 25.

. An overhead transparency would be helpful when discussing the solution.

Day 2. Answers: 6 is opposite 5;
3 is opposite 1;
4 is opposite 2

Comments and suggestions:

. Some pupils may need to use a model of a cube in order to solve the problem.

. Others may see that the "2" is always on top but has been rotated in each drawing.

. Pupils could invent similar dice problems in which faces were numbered differently.

Day 3. Answer: 20 girls

Comments and suggestions:

. Pupils will probably place numbers around a circle in order to solve the problem. You should remind them that the use of a drawing is often very helpful in problem solving.

Make And Use A Drawing Or Model (cont.)

WEEK 4 - DAY 4

How many ways can you buy 4 attached stamps at the post office?
Make drawings to show at least 10 different ways. Two of them
are shown.

****

WEEK 4 - DAY 5

Use bottlecaps or markers
        to make ⤵

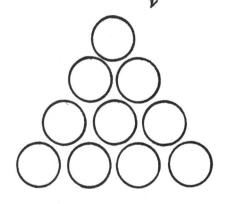

Move <u>only</u> 3 bottlecaps or
    markers to make ⤵

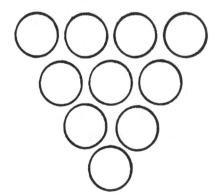

Day 4. Answers: There are 19 possible arrangements.

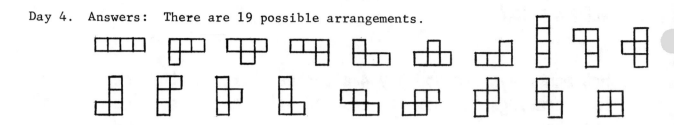

Comments and suggestions:

. Possibly a supply of Green Stamps would help pupils see the possibilities.

. Note that there is a system in the way the drawings of the answer are presented. Perhaps an overhead transparency could be made so that you could emphasize this systematic listing.

Day 5. Answer:

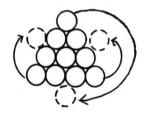

Comments and suggestions:

. You will need 10 bottlecaps, counters, or coins for each pupil (or each pair of pupils). Expect some pupils to have difficulty with this problem.

. You might suggest to those pupils experiencing frustration that that they try a symplified version of the problem:

e.g. move two corner pieces to change the top figure into the bottom figure.

The solution of the easier problem may help them solve the more difficult one.

Eliminate Possibilities

Usually students try to solve a problem by looking directly for the answer, but sometimes it is more helpful to identify or list possible answers and then eliminate incorrect answers. This narrows the search for a correct answer and in some cases leaves only the correct answer.

Ms. Tuel used this problem to introduce <u>eliminate</u> <u>possibilities</u>:

Find the ages of my three children.
Clue 1: The product of the three ages is 36.
Clue 2: Two of the children are twins.
Clue 3: The youngest is not a twin.

Ms. Tuel: Now that we've read the entire problem, let's use just the first two clues and list all the possibilities. We have to have 3 numbers whose product is 36 and two of the numbers are equal. What are some possibilities?

Philip: 4 times 4. No - 2 and 2 and 9.

Jane: How about 3 and 3 and 4.

Ms. Tuel: Maybe I'd better make an organized list on the board.

writes   2 x 2 x 9 = 36
3 x 3 x 4 = 36

Carrie: 4 and 5 don't work. 6, 6 and 1 do.

Greg: So does 1 times 1 times 36.

Ms. Tuel: Is that it? We have four possibilities.   1 x 1 x 36
2 x 2 x 9
Now which ones can we eliminate?   3 x 3 x 4
6 x 6 x 1

Al: The twins can't be youngest so cross off the first three.

Ms. Tuel: That leaves us with the twins as 6 years old and the other child 1 year old. By <u>eliminating</u> <u>possibilities</u> we are left with the answer. <u>Eliminating</u> <u>possibilities</u> is our new problem-solving skill for this week.

ELIMINATE POSSIBILITIES

WEEK 5 - DAY 1

Find the ages of my three children.

Clue 1: The product of the three ages is 36.
Clue 2: Two of the children are twins.
Clue 3: The youngest is not a twin.

****

WEEK 5 - DAY 2

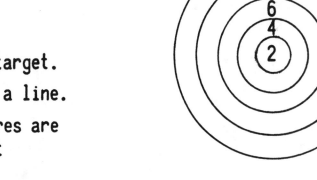

Sue threw 5 darts.

Each dart hit the target.

No darts landed on a line.

Which of these scores are you certain are not possible?

38   25   60   30   42   37   26   8   14

Eliminate Possibilities

Day 1. Answer: 6, 6, 1

          Comments and suggestions:

          . See the introductory commentary on page 31 for suggestions
            on introducing this problem.

Day 2. Answers: 25, 60, 37 and 8

          Comments and suggestions:

          . 25 and 37 are not possible because even numbers can't add
            up to an odd number.

          . 50 is the maximum score and 10 is the minimum score so 60
            and 8 are not possible. The rest are possible.

Eliminate Possibilities (cont.)

WEEK 5 - DAY 3

Ms. Ashley has less than 100 pieces of candy.

If she makes groups of 2 pieces, she will have 1 piece left over.
If she makes groups of 3 pieces, she will have 1 piece left over.
If she makes groups of 4 pieces, she will have 1 piece left over.
If she makes groups of 5 pieces, she will have no pieces left over.

How many pieces of candy could she have?

****

WEEK 5 - DAY 4

Each different letter stands for
a different digit. Find an
addition problem that works.

```
   D O G
 + C A T
 -------
 T O A D
```

****

WEEK 5 - DAY 5

A number times itself is 841.
What is the number?

Day 3. Answers: 25, 85

Comments and suggestions:

. It is helpful to use the last clue first.
  0,5,10,15,20,25,30,35,40,45,50,55,60,65,70,75,80,85,90,95

. The first clue eliminates even numbers and leaves

    5, 15, 25, 35, 45, 55, 65, 75, 85 and 95

. The second clue eliminates numbers that are one more than
  a multiple of 3 and leaves 25, 55, and 85.

. The third clue eliminates 55 and leaves 25 and 85 as the
  two possible answers.

Day 4. Answers:

302	403	605	706	807
7A1	6A1	4A1	3A1	2A1
10A3	10A4	10A6	10A7	10A8

Where A can be any remaining digit.

Comments and suggestions:

. Possibilities for T are 1-9 but 2-9 can be eliminated since
  a 2 or more for T would require 9 - - and a carry of at
  least 3 in the hundreds place.

$$\begin{array}{r} 9\ \text{-}\ \text{-} \\ 8\ \text{-}\ \text{-} \\ \hline 2\text{-}\ \text{-}\ \text{-} \end{array}$$

T must be 1.

. G can't be 1 because 1 is used; G can't be zero because T and D
  can't both be 1. Let G be any other digit, say 5, then D is 6.

. Since 0 + A = A (notice there is no regrouping digit if G = 5),
  0 must be zero. A can be any remaining digit but notice C must
  be 4 to give 605 as one possibility.

$$\begin{array}{r} 605 \\ 491 \\ \hline 1096 \end{array}$$

Day 5. Answer: 29

Comments and suggestions:

. Before any guesses are made, all possibilities less than 30 can
  be eliminated because 30 x 30 = 900.

. Also since the units digit in the product is 1, all factors,
  except those ending in 1 or 9, can be eliminated.

Grade 7

II. DRILL AND PRACTICE – WHOLE NUMBERS

Most seventh-grade classes are a collection of pupils with varying levels of skills. What can be done to provide additional practice and learning for all pupils? One solution is to offer whole number drill and practice through problem-solving activities. While Tim is remembering that 3 x 8 is 24, not 32, Hosea might be figuring out all the possible ways to complete ____ x ____ = 24.

One marvelous discovery in using problem-solving skills with a mixed class is that some pupils who are unskilled in computation are really good at seeing patterns or figuring out ways to solve problems. Some who hate ordinary drill exercises will fill a page with computation to try to solve a problem..

Using The Activities

The activities in this section can be incorporated with the regular teaching and review of whole numbers. Most of them should be done in class so that the teacher can interact and give appropriate hints when necessary. There are many opportunities for pupils to create problems of their own. This type of activity should be encouraged since pupils will become more aware of the real structure of a problem if they have a chance to create one of their own.

You need to be aware that several important properties are developed in some of the lessons (see below). You may wish to provide more practice with these properties and perhaps identify them by name.

"Machine Hook-ups"	. Inverse operations . Addition property of 0 . Multiplication property of 1
"Using Parentheses"	. Associative properties of addition and multiplication
"Shortcuts"	. Distributive property
"Across and Down"	. Rearrangement properties of addition and multiplication
"Addition and Multiplication Tables"	. Addition property of 0 . Multiplication property of 1 . Commutative properties of addition and multiplication.

HUNDRED DOLLAR WORDS

A = $ 1	J = $ 10	S = $ 19
B = $ 2	K = $ 11	T = $ 20
C = $ 3	L = $ 12	U = $ 21
D = $ 4	M = $ 13	V = $ 22
E = $ 5	N = $ 14	W = $ 23
F = $ 6	O = $ 15	X = $ 24
G = $ 7	P = $ 16	Y = $ 25
H = $ 8	Q = $ 17	Z = $ 26
I = $ 9	R = $ 18	

1. Show that "Portland" is a $100 word.

2. How many dollars is "Eugene" worth?

3. How much is your town worth?

4. Who has the most expensive name in the class?

5. Find a 3-letter word that has

    a. the cheapest value.   b. the most expensive value.

6. Find other $100 words besides Portland.

Hundred Dollar Words

Mathematics teaching objectives:

. Develop computational and mental arithmetic skills.

Problem-solving skills pupils _might_ use:

. Guess and check.

. Share data and results with others.

. Search printed matter for needed information (including the use of a dictionary).

Materials needed:

. None

Comments and suggestions:

. Introduce the activity to the whole class allowing time for individual work with frequent opportunities for sharing.

. Letter values might be placed on transparency.

. The only basic strategy for finding $100 words seems to be random guessing with some refining and much brute force.

Answers:

1. Sum of the values for each letter in PORTLAND is $100.

2. Value of EUGENE - $57.

3. & 4. Answers will vary.

5. Some possibilities:

   (a) CAB - $6   (b) WRY - $66
       BAA - $4   TUX - $65
       ABA - $4   ZYM - $64
                              YOU - $61
                              WOW - $61

6. Some possibilities:

   USELESS, WATERGATE, TURKEY, TELEPHONE, HEMOGLOBIN, ELEPHANTS, HOSPITAL, PUMPKIN, STYLES

   Here are some sentences in which each word is $100:

   PREVENT INFLATION!   SUZANNE IMPORTED VIOLINS WHOLESALE.
        FILTERING UPSETS GRUMPY SMOKERS.

Pupils might enjoy making their own extensions.

FINDING PATTERNS

Study this pattern: 3, 4, 7, 11, 18, 29, 47, 76

Note that   3 + 4 = 7
            4 + 7 = 11
            7 + 11 = 18
                 etc.

Use the same rule to complete the patterns below.

1.   1, 1, 2, 3, 5, 8, ___, ___, ___, 55

2.   2, 4, 6, 10, 16, ___, ___, ___, 110

3.   3, 1, 4, 5, 9, 14, ___, ___, ___, 97

Use the same technique as above.

4.   3, 7, ___, ___, ___, ___, 71

5.   1, 5, ___, ___, ___, ___, ___, 73

6.   6, ___, 14, ___, ___, ___, ___, ___, 246

7.   2, ___, 10, ___, ___, ___, ___, ___, 194

8.   6, ___, ___, 12, ___, ___, ___, 87

9.   1, ___, ___, ___, 14, ___, ___, 60

10.   3, ___, ___, ___, 30, ___, ___, 128

BRAIN BUSTERS

11.   6, ___, ___, ___, ___, ___, ___, ___, 183

12.   5, ___, ___, 12, ___, ___, 53

13.   Make up one of your own. See if a friend can solve it.

PSM 81

Finding Patterns

Mathematics teaching objectives:

. Develop mental arithmetic skills.

Problem-solving skills pupils <u>might</u> use:

. Guess and check.

. Identify patterns suggested by data.

. Work backwards.

Materials needed:

. None

Comments and suggestions:

. Inform pupils they are to work as much of the page as they can without help from others. After they have spent 10 minutes or more on the sheet and are "stuck" you might suggest that they get a hint from a classmate. However, this activity has little value if it becomes a cooperative enterprise.

. You should not offer any direct instruction but should observe, listen, and encourage pupils to discuss with you the strategies they are using.

. Problem 12 is impossible if we are restricted to whole numbers. If this restriction is removed, many students should be able to find the solution.

Answers:

1. 1, 1, 2, 3, 5, 8, <u>13</u>, <u>21</u>, <u>34</u>, 55
2. 2, 4, 6, 10, 16, <u>26</u>, <u>42</u>, <u>68</u>, 110
3. 3, 1, 4, 5, 9, 14, <u>23</u>, <u>37</u>, <u>60</u>, 97
4. 3, 7, <u>10</u>, <u>17</u>, <u>27</u>, <u>44</u>, 71
5. 1, 5, <u>6</u>, <u>11</u>, <u>17</u>, <u>28</u>, <u>45</u>, 73
6. 6, <u>8</u>, 14, <u>22</u>, <u>36</u>, <u>58</u>, <u>94</u>, <u>152</u>, 246
7. 2, <u>8</u>, <u>10</u>, <u>18</u>, <u>28</u>, <u>46</u>, <u>74</u>, <u>120</u>, 194
8. 6, <u>3</u>, <u>9</u>, 12, <u>21</u>, <u>33</u>, <u>54</u>, 87
9. 1, <u>4</u>, <u>5</u>, <u>9</u>, 14, <u>23</u>, <u>37</u>, 60
10. 3, <u>8</u>, <u>11</u>, <u>19</u>, 30, <u>49</u>, <u>79</u>, 128
11. 6, <u>5</u>, <u>11</u>, <u>16</u>, <u>27</u>, <u>43</u>, <u>70</u>, <u>113</u>, 183
12. 5, $3\frac{1}{2}$, $8\frac{1}{2}$, 12, $20\frac{1}{2}$, $32\frac{1}{2}$, 53

13. Answers will vary.

MACHINE HOOK-UPS

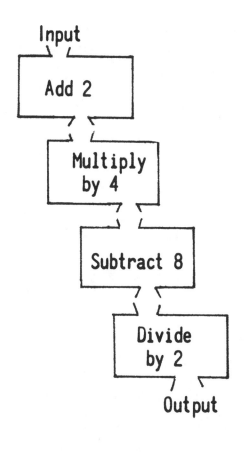

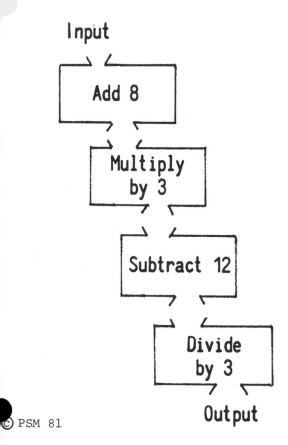

1. a. Try putting 5 into the top machine. What number comes out of the bottom machine? Put your answer in the table. Then complete the rest of the table by putting in other numbers.

Input	Output
5	
9	
0	
20	

   b. What do you think the output will be with an input of 7? First make a guess. Then test it out.

   c. What one machine could be used in place of all 4 machines?

2. a. Use the new hook-up at the left. Complete the following table.

Input	Output
2	
3	
0	
10	

   b. What do you think the output will be with an input of 15? First make a guess. Then test it out.

   c. What one machine could be used in place of all four?

Machine Hook-Ups

Mathematics teaching objectives:

. Provide readiness for working with formulas and equations.

. Use the undoing characteristic of inverse operations.

. Use addition property of zero.

. Use the multiplication property of one.

. Practice computation skills.

Problem-solving skills pupils _might_ use:

. Guess and check.

. Make predictions based upon observed pattern.

. Make a systematic listing or table.

. Create new problems by varying an old one.

Materials needed:

. None

Comments and suggestions:

. Upon completion of this lesson it would be profitable if you discussed the many different strategies used to reach the conclusions.

. This could be a teacher directed activity with an overhead transparency used to show hook-ups.

Answers:

1. a.

Input	Output
5	10
9	18
0	0
20	40

b. Answers will vary. The correct output is 14. A fruitful class discussion topic might be, "What strategies did you use in making your guesses?"

c. Multiply by 2.

2. a.

Input	Output
2	6
3	7
0	4
10	14

b. Answers will vary. The correct output is 19. As before, a fruitful class discussion topic might be, "What strategies did you use in making your guesses?"

c. Add 4

Machine Hook-Ups (cont.)

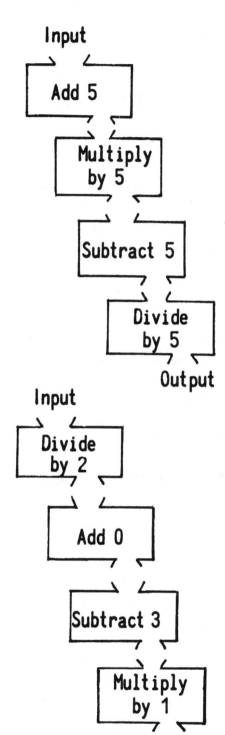

Input

Add 5

Multiply by 5

Subtract 5

Divide by 5

Output

Input

Divide by 2

Add 0

Subtract 3

Multiply by 1

Output

3. a. Here's a 5-5-5-5 hook-up. Put in several numbers to see what happens. What have you discovered?

   b. What will happen in a 6-6-6-6 hook-up? Experiment to find out.

   c. What do you think will happen in a 3-3-3-3 hook-up. First make a guess. Then test it out.

4. a. Notice that this machine hook-up is different. The "divide machine" is first. Use this hook-up to complete the table.

Input	Output
10	
20	
16	
30	

   b. What do you think the output will be with an input of 100? First make a guess. Then test it out.

   c. What two of the machines could be used in place of all four?

   d. What effect does an "add 0" machine have?

   e. What effect does a "multiply by 1" machine have?

Answers: (cont.)

3. In each case, the four machines produce the same result as a single add machine.

    a. A 5-5-5-5 hook-up can be replaced by an add 4 machine.

    b. A 6-6-6-6 hook-up can be replaced by an add 5 machine.

    c. A 3-3-3-3 hook-up can be replaced by an add 2 machine.

4. a.

Input	Output
10	2
20	7
16	5
30	12

b. Answers will vary. Another good class discussion item. The correct output is 47.

c. Divide by 2; subtract 3

d. Output equals input.

e. Output equals input.

Machine Hook-Ups (cont.)

5. Use the drawings below to "build" your own machine hook-ups so that

. in hook-up (a) the output will be the same as the input,
. in hook-up (b) an input of 5 will produce an output of 11,
. in hook-up (c) the output will always be less than the input.

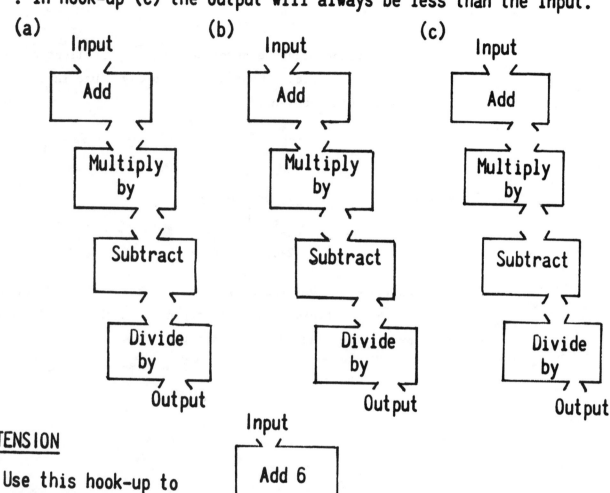

(a) Input → Add → Multiply by → Subtract → Divide by → Output

(b) Input → Add → Multiply by → Subtract → Divide by → Output

(c) Input → Add → Multiply by → Subtract → Divide by → Output

EXTENSION

Use this hook-up to complete the table.

Input	Output
	39
	47
	60

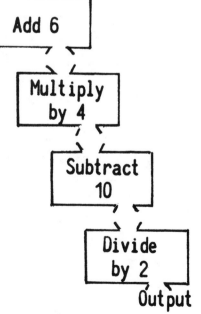

Input → Add 6 → Multiply by 4 → Subtract 10 → Divide by 2 → Output

Answers: (cont.)

5. a. Answers will vary.
    b.   "   "   "
    c.   "   "   "

The exercise might be developed into an excellent class discussion topic. "What strategy did you use in building your machine hook-ups?"

Extension:

Input	Output
16	39
20	47
26.5	60

Pupils might get the answers by guess, check, and refine or by working backwards. This extension could be made into a profitable class discussion topic.

BLANK SQUARES

Complete the multiplication tables by filling in the blanks.

1.

X	5	10	19	6
6	30	60		
25	125			
20				
	45			

2.

X		12	15	
6				
		120		80
9				
	80		300	

3.

X	24		78	
	288			492
32				
		2754	4212	
73				

4.

X						
		40		15		
			40		72	
			15			
	8				12	36
			45			
	14					42

5. Use this blank table to create your own problem. Give just enough clues so that a classmate would be able to complete the table.

X					

Blank Squares

Mathematics teaching objectives:

. Provide practice in multiplication and division.
. Use number theory properties.

Problem-solving skills pupils might use:

. Use a table.
. Work backwards. (use inverse operations)
. Create a problem similar to those already solved.

Materials needed:

. None

Comments and suggestions:

. You may have to review the basic structure of a multiplication chart.

. At some time, either in a discovery session or a culminating sharing period, strategies should be discussed. Different successful strategies of pupils should be highlighted.

. The key strategy for completing the last tables is to recognize that every number in a column must have a common divisor. The same idea applies to each row.

Answers:

1.

X	5	10	19	6
6	30	60	114	36
25	125	250	475	150
20	100	200	380	120
9	45	90	171	54

2.

X	4	12	15	8
6	24	72	90	48
10	40	120	150	80
9	36	108	135	72
20	80	240	300	160

3.

X	24	51	78	41
12	288	612	936	492
32	768	1632	2496	1312
54	1296	2754	4212	2214
73	1752	3723	5694	2993

4.

X	2	8	5	3	9	6
5	10	40	25	15	45	30
8	16	64	40	24	72	48
3	6	24	15	9	27	18
4	8	32	20	12	36	24
9	18	72	45	27	81	54
7	14	56	35	21	63	42

5. Answers will vary.

USING PARENTHESES

1. See if you can score 10 points in this "game of parentheses." Here are the directions:

   - Each problem at the right is done incorrectly.
   - Parentheses can be inserted so that each answer is correct.
   - You score 2 points for each pair of parentheses used correctly.
   - Whenever parentheses are used it means that you do what's inside the parentheses first.

   $5 \times 10 + 1 = 55$

   $2 \times 4 + 3 + 7 = 28$

   $3 \times 4 + 5 + 2 = 29$

   $25 - 15 - 2 = 12$

   $17 - 11 + 3 = 3$

   ## Example

   $2 \times 3 + 4 + 5 = 19$ $2 \times (3 + 4) + 5 = 19$

   Incorrect   Correct

2. Find these answers. Remember, do what's inside the parentheses first.

   a. $15 + (13 + 1) =$
      $(15 + 13) + 1 =$

   e. $20 - (5 - 4) =$
      $(20 - 5) - 4 =$

   b. $17 - (14 - 2) =$
      $(17 - 14) - 2 =$

   f. $36 \div (12 \div 3) =$
      $(36 \div 12) \div 3 =$

   c. $6 \times (3 \times 2) =$
      $(6 \times 3) \times 2 =$

   g. $14 + (6 + 4) =$
      $(14 + 6) + 4 =$

   d. $8 \div (4 \div 2) =$
      $(8 \div 4) \div 2 =$

   h. $2 \times (5 \times 6) =$
      $(2 \times 5) \times 6 =$

Using Parentheses

Mathematics teaching objectives:

. Use the associative properties of addition and multiplication.

. Investigate the order of operations and the use of parentheses.

. Practice computation skills.

Problem-solving skills pupils might use:

. Study the solution process.

. Recognize properties suggested by data.

. Make a general statement concerning an investigation.

Materials needed:

. None

Comments and suggestions:

. You may wish to follow this lesson with a more complete treatment on order of operations.

. In this lesson, the main objective is to show that the order in
which you carry out operations can result in different answers.
Parentheses are introduced as a way of indicating when the operations are to be done in a different order than from left to right.

Answers:

1. $5 \times (10 + 1) = 55$ $25 - (15 - 2) = 12$

$2 \times (4 + 3 + 7) = 28$ $17 - (11 + 3) = 3$

$3 \times (4 + 5) + 2 = 29$

2. a. 29, 29 (addition)   e. 19, 11 (subtraction)

b. 5, 1   (subtraction)   f. 9, 1   (division)

c. 36, 36 (multiplication)   g. 24, 24 (addition)

d. 4, 1   (division)   h. 60, 60 (multiplication)

Using Parentheses (cont.)

3. Study the problems and answers in exercise 2. Sometimes
   the answers in the oval are the same; sometimes they're
   different. When does this seem to happen?

4. Study the problems below. First try to predict whether the
   answer in the oval will be the same or different. Then work
   the problems to check your predictions.

$$121 - (23 - 18) =$$
$$(121 - 23) - 18 =$$

$$7 \times (8 \times 11) =$$
$$(7 \times 8) \times 11 =$$

$$139 + (61 + 39) =$$
$$(139 + 61) + 39 =$$

$$144 \div (12 \div 4) =$$
$$(144 \div 12) \div 4 =$$

5. Write a sentence or two about your conclusions.

EXTENSION

Insert parentheses to make this statement true. More than
one set of parentheses may be necessary.

$$5 \times 3 + 6 + 10 - 5 - 3 + 2 = 95$$

Answers: (cont.)

3. When the exercises involve multiplication or addition, the answers are the same. When exercises involve subtraction and division, the answers are different.

4. Different   116, 80
   Same   616, 616
   Same   239, 239
   Different   48, 3

5. If two addition operations are used, the answers are the same regardless of the order in which the operations are performed. The same is true for multiplication. This is not true for subtraction and division.

Extension: One possible answer:

$$5 \times (3 + 6 + 10) - \left[5 - (3 + 2)\right] = 95$$

DIGIT DRAW ACTIVITIES
(Ideas for Teachers)

Ten digit cards marked 0-9 can be used
for a variety of activities.

A. Have pupils make the diagram to the right.

B. Suggest a goal, such as, getting the
   largest possible sum.

C. Shuffle and draw digits, one at a time. Pupils must write
   each digit, as it is drawn, until all spaces are filled.

D. Compare and discuss results.

This activity is very adaptable. For example, in the 3-digit by
3-digit addition problem above, pupils may not get the largest
sum the first time. The teacher can have the pupils use the same
six digits to find the largest sum; then use the same six digits
to find the smallest sum; and then use those digits to find the
sum closest to, say, 700.

These activities can be used for drill and practice, concept
development and/or diagnosis. Each is highly individual, as many
pupils will have a unique problem. See the next page for other
suggested formats. Further variations could include:

   . replacing a drawn digit so it can be used again.

   . restricting certain numbers, e.g., no zero is used
     with division and fraction problems.

   . allowing a special reject box giving pupils the chance
     to discard an unfavorable draw.

   . adapting whole number activities to decimal activities
     by inserting a decimal point(s).

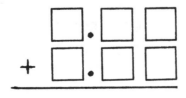

<u>Digit</u> <u>Draw</u> <u>Activities</u>

Mathematics teaching objectives:
- . Develop place value concepts.
- . Develop informal probability concepts.
- . Practice computation skills.

Problem-solving skills pupils <u>might</u> use:
- . Break problem into manageable parts.
- . Make decisions based upon data.
- . Recognize limits and/or eliminate possibilities.

Materials needed:
- . Digit cards 0-9

Comments and suggestions:
- . This activity works well as an opener at the beginning of the class
  or as an ending for that last few minutes of a class when all other
  activities have been completed.

- . After some trials, pupils can be encouraged to share what strategies they use. "If a 9 or 8 is drawn first, put it in the 100's
  place. If a little digit comes first, put it in the 1's place.
  It's hard to decide what to do with a 5 or 6. Sometimes it's
  just luck!"
  The strategies pupils use to solve problem 3 on the second page
  will vary greatly. One pupil might write out lots of possibilities
  to see which one gives smaller answers. This pupil needs en-
  couragement in organizing the trials and making conclusions from
  the trials that give larger answers. Another might try to ana-
  lyze the problem seeing into the subtraction process. Others
  might take the problem to their parents or an older friend for
  some shared problem solving. Pupils can be complimented on their
  efforts even if they don't find the smallest difference. The
  smallest difference found could be posted and changed as someone
  finds a smaller difference.

Digit Draw Activities (cont.)

OTHER FORMATS

Place Value: □,□□□   largest number, smallest number,
or closest to 5000

Ordering: □ □ < □ □ < □ □

Addition: □ □   □□□   Subtraction: □□□   □□□
□ □ +□□□   − □□   − □□□
+ □ □

Multiplication: □□□   □□□□   □□   □□□
× □   × □   ×□□   ×□□

Division: □⟌□□□   □⟌□□□□   □□⟌□□□□

Fractions: □/□ + □/□ or any other operation

Two other activities, seemingly obvious, lead to an interesting third problem.

1. Using each digit once, make the largest possible 10-digit number.

2. Using each digit once, make the largest possible sum for a 5-digit by 5-digit addition problem.

3. Using each digit once, make the smallest possible difference for a 5-digit by 5-digit subtraction problem.

The answer to (1) is 9,876,543,210. The answer to (2) is 97,531 + 86,420 = 183,951. Many pupils will think the answer to (3) is 97,531 − 86,420 = 11,111. But a much smaller difference is possible--50,123 − 49,876 = 247.

CREATE A PROBLEM

Write a single digit in each square to create a correct problem.
Digits may be repeated in a problem.

1. ☐ ☐ ☐
    + ☐ ☐ ☐
    ———
    5 6 1

2. ☐ ☐ ☐
    + 8 7 3
    ———
    ☐ ☐ ☐ ☐

3. ☐ 9
    7 ☐
    + ☐ ☐
    ———
    2 ☐ ☐

4. ☐ ☐ ☐
    − 5 8 9
    ———
    ☐ ☐ ☐

5. 5 2 8 4
    − 9 ☐ ☐
    ———
    ☐ ☐ ☐ ☐

6. ☐ ☐ ☐ ☐
    − ☐ 4 ☐
    ———
    2 9 3

7. ☐ ☐
    × ☐ 8
    ———
    ☐ ☐ ☐

8. 1 9 5
    × ☐ ☐
    ———
    ☐ ☐ ☐ ☐ ☐

9. ☐ ☐ ☐
    ×   ☐
    ———
    3 3 7 2

10. ☐ ☐ ☐
    × 2 9
    ———
    ☐ ☐ ☐ ☐ 1

11. ＿＿2 7＿
    ☐ / ☐ ☐ ☐

12. ＿1 2 3 4＿ R 5
    ☐ / ☐ ☐ ☐ ☐

13. ＿＿☐ ☐ ☐＿
    5 3 / ☐ ☐ ☐ ☐

14. ＿＿☐ ☐ ☐＿
    ☐ / 1 2 3 4

15. ＿＿2 3＿ R 31
    ☐ ☐ / ☐ ☐ ☐ ☐

Create your own problems. Give them to a friend to solve.

Create A Problem

Mathematics teaching objectives:
. Practice computation skills.
. Use inverse operations.

Problem-solving skills pupils might use:
. Work backwards (use inverse operations).
. Recognize limits and/or eliminate possibilities.
. Guess and check.

Materials needed:
. Calculators (optional)

Comments and suggestions:
. Encourage pupils not to use the obvious answers involving zeros and ones.
. Some pupils will solve problems like 9 and 12 easily by using inverse operations. Others will use a guess, check and refine method. Both ways yield beneficial results.
. Some pupils could determine the limits for certain numbers. For example, the top number in problem 7 can range from 13 to 99.
. This activity is difficult to score since each pupil might have unique answers. An alternative to teacher scoring is to provide calculators for pupils to check their own or a classmate's paper. You can then look at incorrect solutions to diagnose troubles with with computation skills, regrouping skills, etc.

Answers:
Answers will vary. Some possibilities are given.

1. $349 + 212 = 561$
2. $356 + 873 = 1229$
3. $89 + 75 + 76 = 240$
4. $714 - 589 = 125$
5. $5284 - 946 = 4338$
6. $1041 - 748 = 293$
7. $36 \times 8 = 288$
8. $195 \times 83 = 16185$
9. Only two solutions are possible:
   $562 \times 6 = 3372$
   $843 \times 4 = 3372$

10. $789 \times 29 = 22881$
11. $162 \div 6 = 27$
12. $7409 \div 6 = 1234 \text{ R } 5$
13. $7155 \div 53 = 135$
14. Only one solution is possible:
    $1234 \div 2 = 617$
15. $1066 \div 45 = 23 \text{ R } 31$

-62-

SHORTCUTS

1. a. $(7 \times 98) + (7 \times 2) = \boxed{}$

    b. $(9 \times 97) + (9 \times 3) = \boxed{}$

    c. $(5 \times 93) + (5 \times 7) = \boxed{}$

    d. $(14 \times 88) + (14 \times 12) = \boxed{}$

    e. $(23 \times 95) + (23 \times 5) = \boxed{}$

    f. $(31 \times 25) + (31 \times 75) = \boxed{}$

2. Todd and Tim were arguing about the answer to this problem.
    Todd said the answer was 2600.
    Tim said he didn't know what the answer was--but he was certain it was less than 2600.
    Without working the problem, how can you tell which boy is correct?

    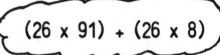

    $(26 \times 91) + (26 \times 8)$

3. Examine the problems below. Decide which of them can be worked mentally. Then solve only those that can be worked mentally.

    a. $(17 \times 90) + (17 \times 10)$ d. $(42 \times 75) + (42 \times 25)$

    b. $(26 \times 91) + (26 \times 9)$ e. $(12 \times 85) + (12 \times 3)$

    c. $(15 \times 93) + (15 \times 5)$ f. $(29 \times 60) + (29 \times 40)$

4. For each problem below, the same number should be placed in both boxes.

    a. $(\boxed{} \times 96) + (\boxed{} \times 4) = 2500$

    b. $(\boxed{} \times 89) + (\boxed{} \times 11) = 3000$

    c. $(\boxed{} \times 85) + (\boxed{} \times 15) = 5000$

Shortcuts

Mathematics teaching objectives:
. Use the distributive property.
. Practice computation skills.

Problem-solving skills pupils might use:
. Recognize properties suggested by data.
. Study the solution process.
. Guess and check.

Materials needed:
. None

Comments and suggestions:
. This lesson should be done during class so that you can interact.
. You may wish to provide more practice and eventually formalize
the distributive property: $a(b + c) = ab + ac$

Answers:

1. a. 700
    b. 900
    c. 500   Hopefully, pupils will discover that in
    d. 1400   each case the sum of the 2nd and 4th
    e. 2300   numbers is 100.
    f. 3100

2. Tim is correct. The problem can be worked by adding 91 and 8
    and then multiplying by 26. This is less than 2600.

3. Those that can be worked mentally:
    a. 1700   d. 4200
    b. 2600   f. 2900

4. a. 25
    b. 30
    c. 50

Shortcuts (cont.)

5. Solve each problem. Be sure to work them out completely. You may be surprised with the results.

   a. $(51 \times 7) + (51 \times 3) = \boxed{}$

   b. $(73 \times 4) + (73 \times 6) = \boxed{}$

   c. $(59 \times 8) + (59 \times 2) = \boxed{}$

   d. $(88 \times 9) + (88 \times 1) = \boxed{}$

6. For each problem below the two numbers in the boxes should be different.

   a. $(35 \times \boxed{}) + (35 \times \boxed{}) = 350$

   b. $(46 \times \boxed{}) + (46 \times \boxed{}) = 460$

   c. $(21 \times \boxed{}) + (21 \times \boxed{}) = 2100$

   d. $(18 \times \boxed{}) + (18 \times \boxed{}) = 1800$

7. You've discovered that

   $(18 \times 93) + (18 \times 7)$ has the same result as $18 \times (93 + 7)$.

   Each problem in exercise 1 also can be rewritten in a similar way. Show how this can be done.

   Example: a. $(7 \times 98) + (7 \times 2) = 7 \times (98 + 2)$.

Answers: (cont.)

5.   a.   510

     b.   730

     c.   590

     d.   880

6.   Answers will vary.   Some possibilities are given:

     a.   8, 2

     b.   7, 3

     c.   97, 3

     d.   80, 20

7.   a.   (7 x 98) + (7 x 2) = 7 x (98 + 2)

     b.   (9 x 97) + (9 x 3) = 9 x (97 + 3)

     c.   (5 x 93) + (5 x 7) = 5 x (93 + 7)

     d.   (14 x 88) + (14 x 12) = 14 x (88 + 12)

     e.   (23 x 95) + (23 x 5) = 23 x (95 + 5)

     f.   (31 x 25) + (31 x 75) = 31 x (25 + 75)

ACROSS AND DOWN

When Jill came to class she found this problem on the board.

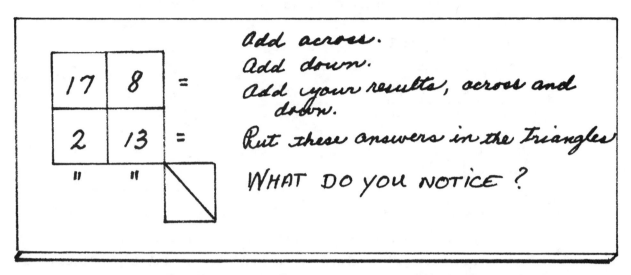

Add across.
Add down.
Add your results, across and down.
Put these answers in the triangles
WHAT DO YOU NOTICE?

Jill was convinced that this was a very special case. She was sure the teacher had used certain numbers to make it work. So Jill made up one of her own. Try this one to see if it works out.

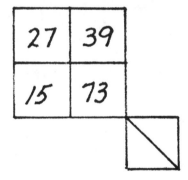

Make up some of your own.

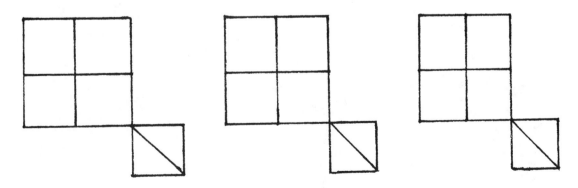

Mathematics teaching objectives:

. Use the rearrangement properties of addition and multiplication.

. Practice computation skills.

Problem-solving skills pupils might use:

. Recognize attributes of tables.

. Create new problems by varying an old one.

. Make explanations based upon data.

Materials needed:

. None

Comments and suggestions:

. Usually several pupils in the class will create problems that "don't work." Perhaps it would be beneficial if their classmates helped them find their mistakes.

. Upon completion of the lesson, you might provide some exercises in which the rearrangement property can be used to simplify computation, e.g., $79 + 86 + 52 + 21 + 14 = (79 + 21) + (86 + 14) + 52$

Answers: (See next page)

Across and Down (cont.)

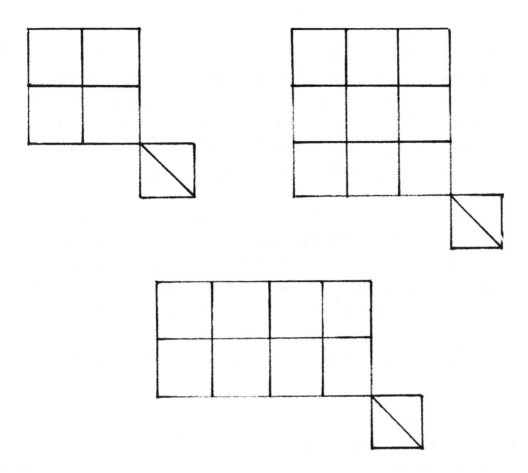

1. Why do you think these problems work out the way they do?

2. Go back and put "ears" on each figure like this.

   In each case, add the numbers diagonally. Put your answers in the ears.

   What do you notice? Explain why some cases work out and some do not.

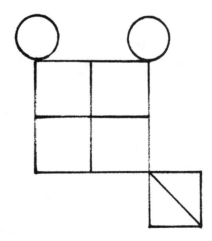

3. Suppose you used multiplication rather than addition. What do you think would happen? You might want to try some cases to see.

Answers:

1. All the examples merely represent different orders used in adding the same given numbers. This is an example of the rearrangement property of addition. One way to explain this is to introduce grouping symbols. For example, the first exercise shows that (17 + 8) + (2 + 13) gives the same results as (17 + 2) + (8 + 13).

2. In the 2 by 2 arrangements, the sum of the numbers in the ears is the same as the numbers in both parts of the split box. This scheme usually does not work for 3 by 3 arrangements since some numbers will not be added and the middle number will be added twice. Students could be challenged to find situations in which the sum of the ears in a 3 by 3 is the same as the numbers in the box.

3. Problems involving multiplication "behave" the same as those with addition.

1. Some tables are easy to complete. Show how to complete each of the tables below.

+	5	11	0	31	27
5	10		5		
11					38
0					
31		42			
27					

x	3	5	10	1	7
3		15			
5					
10			100		
1					
7					

2. Some tables are more challenging.

    Show how to complete this one.

    Use the same numbers across the top as you use on the left side.

    Keep the order of these numbers the same also.

+					
					100
		0		66	
				79	
		91			

3. Now try this one. Use the same rules as for the one above.

x				5	
				35	
		16			
			1		
					65

Addition And Multiplication Tables

Mathematics teaching objectives:

  . Use the addition property of zero.

  . Use the multiplication property of one.

  . Use the commutative properties of addition and multiplication.

Problem-solving skills pupils *might* use:

  . Use a table.

  . Break a problem into parts.

  . Work backwards.

  . Identify patterns suggested by a table.

Materials needed:

  . None

Comments and suggestions:

  . Exercises 1, 2, and 3 probably should be done during class time so you can offer individual assistance when necessary. Pupils need to read carefully the directions for 2 and 3.

  . Upon completion of the lesson, the teacher should point out which properties are used in each exercise.

Answers:

1.

+	5	11	0	31	27
5	10	16	5	36	32
11	16	22	11	42	38
0	5	11	0	31	27
31	36	42	31	62	58
27	32	38	27	58	54

x	3	5	10	1	7
3	9	15	30	3	21
5	15	25	50	5	35
10	30	50	100	10	70
1	3	5	10	1	7
7	21	35	70	7	49

2.

+	9	0	13	66	91
9	18	9	22	75	100
0	9	0	13	66	91
13	22	13	26	79	104
66	75	66	79	132	157
91	100	91	104	157	182

The key to the solution is "0" in the given table. This provides an opportunity to discuss the addition property of zero.

3.

x	7	4	1	5	13
7	49	28	7	35	91
4	28	16	4	20	52
1	7	4	1	5	13
5	35	20	5	25	65
13	91	52	13	65	169

The key to the solution is "1" in the given table. This provides an opportunity to discuss the multiplication property of one.

(Answers continued on next page...)

4. Jeff completed both of the following tables in less than 3 minutes, and he didn't use a calculator. Can you beat Jeff's record? Try it.

+	108	$43\frac{1}{2}$	95	387	792
108	216		2 03		900
$43\frac{1}{2}$	$151\frac{1}{2}$	87		$430\frac{1}{2}$	
95		$138\frac{1}{2}$	190		887
387	495		482	774	
792		$835\frac{1}{2}$		1179	1584

x	35	19	17	40	23
23			391	920	529
17			287	680	
40	1400	760		1600	
35	1225		595		805
19	665	361	323		437

5. The following table is an ordinary multiplication table.

   a. Fill in only the part of the table **below** the dotted line.

x	1	2	3	4	5	6	7	8	9
1									
2									
3									
4									
5									
6									
7									
8									
9									

   b. Now fill in the section of the table above the dotted line. As you fill in this part, look for patterns. Write about anything you discover.

   c. Finally, fill in the numbers on the dotted line.

Answers: (cont.)

4.

+	108	43½	95	387	792
108	216	151½	203	495	900
43½	151½	87	138½	430½	835½
95	203	138½	190	482	887
387	495	430½	482	774	1179
792	900	835½	887	1179	1584

This table will be easy to complete if pupils
have observed the patterns in exercises 1, 2,
and 3. No calculations are necessary. The
exercise provides an opportunity to discuss
the commutative property.

x	35	19	17	40	23
23	805	437	391	920	529
17	595	323	287	680	391
40	1400	760	680	1600	920
35	1225	665	595	1400	805
19	665	361	323	760	437

This table cannot be filled out by using the
symmetric pattern in the other tables. However,
the commutative property still can be used.
It still can be filled out without making any
calculations.

5. This is an ordinary multiplication chart. Pupils are encouraged to
apply the symmetric pattern usually present in addition and multiplication
charts. This symmetry is a consequence of the commutative property.

Grade 7

III. DRILL AND PRACTICE — FRACTIONS

Most seventh-grade classes
are a collection of pupils with
varying levels of skills. What
can be done to provide additional
practice and learning for all
pupils? One solution is to offer
drill and practice with fractions
through problem-solving activities.
While Jim is cutting up circles
to show that $\frac{1}{2}$ is greater than $\frac{1}{3}$,
Jill may be challenged to find
as many fractions between $\frac{1}{2}$ and
$\frac{1}{3}$ as she can.

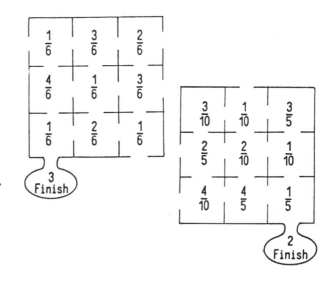

One marvelous discovery in using problem-solving skills with a mixed
class is that some pupils who are unskilled in computation are really good
at seeing patterns or figuring out ways to solve problems. Some who hate
ordinary drill problems will fill a page with computation to try to solve
a problem.

Using The Activities

The activities in this section can be incorporated with the regular
teaching and review of fractions. Most of them should be done in class
so that the teacher can interact and give appropriate hints when necessary.

JUMPING FLOOZIE

Floozie loves to jump. Sometimes he makes little
jumps, sometimes big jumps, sometimes just medium-
sized jumps. Most of the time he jumps forward,
but sometimes he likes to jump backwards. As
long as Floozie is jumping, he's happy.

1. The dots on the lines below show where Floozie landed.
   Write the fraction for each letter.

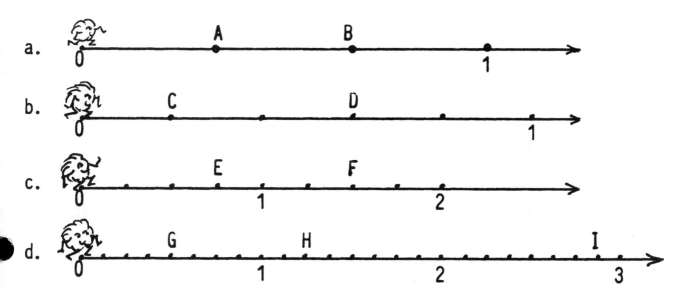

2. On the lines below, Floozie starts off with a jump. On his
   next jump he wants to land on the dot where 1 should be.
   In each case show where 1 should be located.

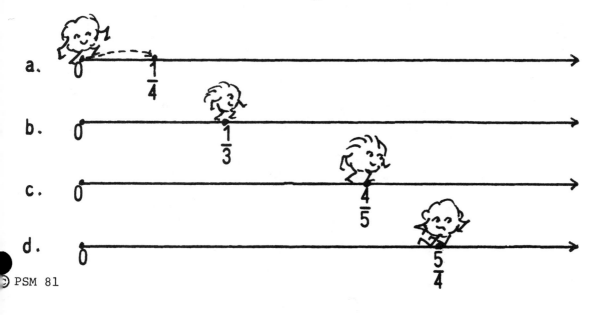

<u>Jumping Floozie</u>

Mathematics teaching objectives:

. Given the zero point and the number for another point on the
  number line, locate other whole numbers and fractions.

Problem-solving skills pupils <u>might</u> use:

. Make and/or use a diagram.

. Guess and check.

. Make decisions based on data.

Materials needed:

. A centimetre ruler (optional)

Comments and suggestions:

. This activity might serve as an assessment instrument to gain insight
  into your pupils' understanding of fractions. Pupils who successfully
  complete this lesson will have demonstrated that they can label points
  on a number line when the unit length is given, and determine the unit
  point when a given point is named with a fraction or mixed number.

. Pupils may need some help on exercises 2 and 3. A guess, check, and
  refine procedure could be suggested. For example, in 2(c), pupils
  could guess the location of $\frac{1}{5}$ and then check this guess by seeing if
  four of these end up at $\frac{4}{5}$.

. Pupils could also use a centimetre ruler to locate the required points.
  However, for some pupils this may cause confusion. (For example, in
  2(b) a matching of "$\frac{1}{3}$" to "4" is required.)

. During discussion be certain that pupils realize that the unit segment
  is not always the same but often varies from one situation to the next.

Answers:

1. A = $\frac{1}{3}$;   B = $\frac{2}{3}$;   C = $\frac{1}{5}$;   D = $\frac{3}{5}$;   E = $\frac{3}{4}$;   F = $1\frac{2}{4}$ or $1\frac{1}{2}$ or $\frac{6}{4}$ or $\frac{3}{2}$

   G = $\frac{4}{8}$ or $\frac{1}{2}$;   H = $1\frac{2}{8}$ or $\frac{10}{8}$ or $1\frac{1}{4}$ or $\frac{5}{4}$;   I = $2\frac{7}{8}$ or $\frac{23}{8}$.

2. Unit segment lengths in centimetres - a. 8 cm   c. 10 cm
                                           b. 12 cm   d. 8 cm

3. The length from point zero to the point labeled 3 -
    a. 12 cm   b. 9 cm   c. 12 cm

3. On the lines below, Floozie wants to land on the dot where
    <u>3</u> should be. In each case, show where <u>3</u> should be located.

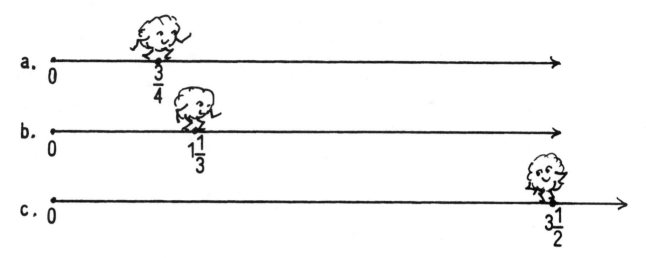

a. 0 $\frac{3}{4}$

b. 0 $1\frac{1}{3}$

c. 0 $3\frac{1}{2}$

4. Use the number lines below.
    Make up and solve your own "Jumping Floozie" problems.

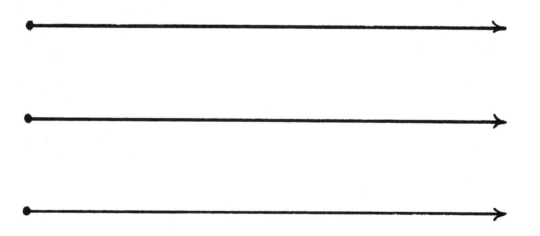

FRACTION PATTERNS

Continue these patterns. Be sure to check to see if you agree with the last number given.

1. 0 $\dfrac{1}{4}$ $\dfrac{2}{4}$ $\dfrac{3}{4}$ ___ ___ ___ ___ $\dfrac{8}{4}$

2. 0 $\dfrac{2}{3}$ $\dfrac{4}{3}$ $\dfrac{6}{3}$ ___ ___ ___ ___ $\dfrac{16}{3}$

3. $\dfrac{24}{4}$ $\dfrac{21}{4}$ $\dfrac{18}{4}$ $\dfrac{15}{4}$ ___ ___ ___ ___ $\dfrac{0}{4}$

4. 12 $10\dfrac{1}{2}$ 9 $7\dfrac{1}{2}$ ___ ___ ___ ___ 0

5. $1\dfrac{2}{5}$ $2\dfrac{4}{5}$ $4\dfrac{1}{5}$ $5\dfrac{3}{5}$ ___ ___ ___ ___ $12\dfrac{3}{5}$

6. $3\dfrac{2}{4}$ $6\dfrac{3}{4}$ 10 $13\dfrac{1}{4}$ ___ ___ ___ ___ $29\dfrac{2}{4}$

7. $1\dfrac{1}{4}$ $2\dfrac{1}{2}$ $3\dfrac{3}{4}$ 5 ___ ___ ___ ___ $11\dfrac{1}{4}$

8. 15 $14\dfrac{5}{8}$ $14\dfrac{1}{4}$ $13\dfrac{7}{8}$ ___ ___ ___ ___ 12

9. 10 5 $2\dfrac{1}{2}$ $1\dfrac{1}{4}$ ___ ___ ___ ___ $\dfrac{5}{128}$

10. $\dfrac{1}{3}$ $\dfrac{2}{3}$ $1\dfrac{1}{3}$ $2\dfrac{2}{3}$ ___ ___ ___ ___ $85\dfrac{1}{3}$

*11. $5\dfrac{1}{8}$ $6\dfrac{3}{4}$ $8\dfrac{3}{8}$ 10 ___ ___ ___ $16\dfrac{1}{2}$

*12. $\dfrac{3}{4}$ $1\dfrac{1}{2}$ $4\dfrac{1}{2}$ 18 ___ ___ ___ 30240

Fraction Patterns

Mathematics teaching objectives:

. Determine operation used on one fraction or mixed number to get another fraction or mixed number.

. Add, subtract, multiply with fractions and mixed numbers.

Problem-solving skills pupils _might_ use:

. Identify patterns suggested by ordered data.

. Make predictions and conjectures based upon observed patterns.

Materials needed:

. None

Comments and suggestions:

. This activity also could serve as an assessment instrument. Pupils who work the exercises readily have a good formal understanding of fractions.

. You may wish to give this activity to pupils with better understanding while you work with others who need a more concrete approach to fractions.

. Some pupils will solve some problems using whole number patterns, e.g., the top numbers in #1 are increasing by 1.

. After pupils finish the activity, they could check their answers with a classmate.

Answers:

1. $\frac{4}{4}$ $\quad$ $\frac{5}{4}$ $\quad$ $\frac{6}{4}$ $\quad$ $\frac{7}{4}$ $\qquad$ 6. $16\frac{2}{4}$ $\quad$ $19\frac{3}{4}$ $\quad$ 23 $\quad$ $26\frac{1}{4}$

2. $\frac{8}{3}$ $\quad$ $\frac{10}{3}$ $\quad$ $\frac{12}{3}$ $\quad$ $\frac{14}{3}$ $\qquad$ 7. $6\frac{1}{4}$ $\quad$ $7\frac{1}{2}$ $\quad$ $8\frac{3}{4}$ $\quad$ 10

3. $\frac{12}{4}$ $\quad$ $\frac{9}{4}$ $\quad$ $\frac{6}{4}$ $\quad$ $\frac{3}{4}$ $\qquad$ 8. $13\frac{1}{2}$ $\quad$ $13\frac{1}{8}$ $\quad$ $12\frac{3}{4}$ $\quad$ $12\frac{3}{8}$

4. 6 $\quad$ $4\frac{1}{2}$ $\quad$ 3 $\quad$ $1\frac{1}{2}$ $\qquad$ 9. $\frac{5}{8}$ $\quad$ $\frac{5}{16}$ $\quad$ $\frac{5}{32}$ $\quad$ $\frac{5}{64}$

5. 7 $\quad$ $8\frac{2}{5}$ $\quad$ $9\frac{4}{5}$ $\quad$ $11\frac{1}{5}$ $\qquad$ 10. $5\frac{1}{3}$ $\quad$ $10\frac{2}{3}$ $\quad$ $21\frac{1}{3}$ $\quad$ $42\frac{2}{3}$

*11. $11\frac{5}{8}$ $\quad$ $13\frac{1}{4}$ $\quad$ $14\frac{7}{8}$

*12. 90 $\quad$ 540 $\quad$ 3780 (Note: The first number is multiplied by 2, the second by 3, the third by 4, etc.)

SQUARES AND PATHS

Find a path that adds up to the number at the Finish. You may enter any open gate.

1.

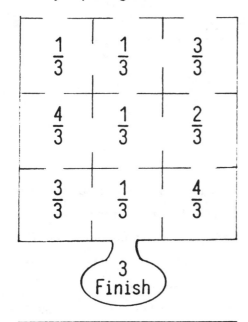

$\frac{1}{3}$	$\frac{1}{3}$	$\frac{3}{3}$
$\frac{4}{3}$	$\frac{1}{3}$	$\frac{2}{3}$
$\frac{3}{3}$	$\frac{1}{3}$	$\frac{4}{3}$

$\frac{3}{}$ Finish

2.

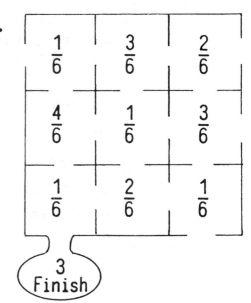

$\frac{1}{6}$	$\frac{3}{6}$	$\frac{2}{6}$
$\frac{4}{6}$	$\frac{1}{6}$	$\frac{3}{6}$
$\frac{1}{6}$	$\frac{2}{6}$	$\frac{1}{6}$

$\frac{3}{}$ Finish

3.

$\frac{1}{2}$	$\frac{3}{4}$	$\frac{1}{2}$
$\frac{1}{4}$	$\frac{2}{4}$	$\frac{2}{2}$
$\frac{3}{2}$	$\frac{1}{4}$	$\frac{6}{4}$

6 Finish

4.

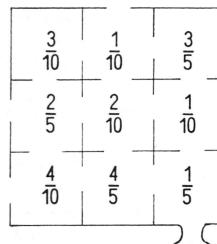

$\frac{3}{10}$	$\frac{1}{10}$	$\frac{3}{5}$
$\frac{2}{5}$	$\frac{2}{10}$	$\frac{1}{10}$
$\frac{4}{10}$	$\frac{4}{5}$	$\frac{1}{5}$

2 Finish

5.

$\frac{1}{2}$	$\frac{3}{6}$	$\frac{2}{3}$
$\frac{1}{6}$	$\frac{2}{2}$	$\frac{4}{6}$
$\frac{1}{3}$	$\frac{1}{6}$	$\frac{1}{2}$

$\frac{3}{}$ Finish

Squares and Paths

Mathematics teaching objectives:

. Add and subtract with fractions.

Problem-solving skills pupils <u>might</u> use:

. Change the problem into an equivalent easier one.
. Guess and check.
. Work backwards.

Materials needed:

. None

Comments and suggestions:

. The first two problems are easy to work since all of the denominators
in the squares are the same. If the "finish" number also is changed to
have the same denominator, only whole-number computation is involved.
The other three problems will be easier if the fractions are changed
to equivalent ones with the same denominator.

. This activity clearly illustrates the skill of "change a problem
into an equivalent easier one " (changing to common denominators).

. This important problem-solving skill needs to be emphasized sometime
during the activity.

Answers: Answers may vary. Here are some possibilities:

1. 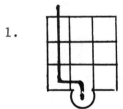   2. 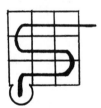   3.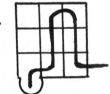

4. 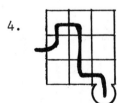   5.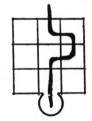

CREATING FRACTION PROBLEMS

Write a fraction or mixed number in each box to create a correct problem.

1. $\boxed{} + \boxed{} = \dfrac{5}{8}$

2. $\dfrac{2}{3} + \boxed{} = \boxed{}$

3. $\boxed{} + \boxed{} = 1$

4. $4\dfrac{1}{2} + \boxed{} = \boxed{}$

5. $\boxed{} + \boxed{} + \boxed{} = 5\dfrac{1}{2}$

6. $\boxed{} - \boxed{} = \dfrac{2}{3}$

7. $\dfrac{5}{6} - \boxed{} = \boxed{}$

8. $\boxed{} - 1\dfrac{1}{3} = \boxed{}$

9. $10 - \boxed{} = \boxed{}$

10. $\boxed{} - \boxed{} = 5\dfrac{1}{4}$

Creating Fraction Problems

Mathematics teaching objectives:

. Practice addition and subtraction of fractions.

. Use inverse operations.

Problem-solving skills pupils might use:

. Work backwards (use inverse operations).

. Recognize limits and/or eliminate possibilities.

Materials needed:

. None

Comments and suggestions:

. Remind pupils that only fractions or mixed numbers are to be used in the boxes.

. Note that there are an unlimited number of possible problems for 1 to 10, but problems 11 to 15 have unique solutions. In order to check the problems created for 1 to 10 you might have the pupils exchange papers and work each others problems.

Answers:

Answers for 1 to 10 will vary. Some possibilities are given.

1. $\frac{2}{8} + \frac{3}{8} = \frac{5}{8}$

2. $\frac{2}{3} + \frac{3}{4} = 1\frac{5}{12}$

3. $\frac{5}{7} + \frac{2}{7} = 1$

4. $4\frac{1}{2} + \frac{3}{4} = 5\frac{1}{4}$

5. $1\frac{1}{2} + 2\frac{1}{2} + 1\frac{1}{2} = 5\frac{1}{2}$

6. $1\frac{1}{3} - \frac{2}{3} = \frac{2}{3}$

7. $\frac{5}{6} - \frac{1}{3} = \frac{1}{2}$

8. $3\frac{2}{3} - 1\frac{1}{3} = 2\frac{1}{3}$

9. $10 - 3\frac{1}{2} = 6\frac{1}{2}$

10. $6\frac{1}{2} - 1\frac{1}{4} = 5\frac{1}{4}$

Creating Fraction Problems (cont.)

11.

$$\frac{2}{3} + \boxed{} = \frac{5}{6}$$

12.

$$\boxed{} + 2\frac{3}{4} = 4$$

13.

$$3 - \boxed{} = 2\frac{1}{3}$$

14.

$$\boxed{} - 1\frac{1}{8} = \frac{5}{8}$$

15.

$$2\frac{1}{2} + 3\frac{1}{2} + 4\frac{1}{2} + \boxed{} = 15$$

*16.   Find three different fractions whose sum is 1.

Creating Fraction Problems

Answers: (cont.)

11. $\frac{2}{3} + \frac{1}{6} = \frac{5}{6}$

12. $1\frac{1}{4} + 2\frac{3}{4} = 4$

13. $3 - \frac{2}{3} = 2\frac{1}{3}$

14. $1\frac{3}{4} - 1\frac{1}{8} = \frac{5}{8}$

15. $2\frac{1}{2} + 3\frac{1}{2} + 4\frac{1}{2} + 4\frac{1}{2} = 15$

*16. One possible answer:

$\frac{1}{3} + \frac{1}{2} + \frac{1}{6} = 1$

SMALLEST ANSWER

Use only the numbers given.
Use each of them only once.
Place them in the boxes to get the smallest possible answer.

1. $\dfrac{\Box}{\Box} + \dfrac{\Box}{\Box} =$

Use
1, 2, 3, 4

2. $\dfrac{\Box}{\Box} - \dfrac{\Box}{\Box} =$

Use
1, 2, 3, 4

3. $\dfrac{\Box}{\Box} \times \dfrac{\Box}{\Box} =$

Use
1, 2, 3, 4

4. $\dfrac{\Box}{\Box} + \dfrac{\Box}{\Box} =$

Use
1, 2, 3, 4

Smallest Answer

Mathematics teaching objectives:

   . Practice addition, subtraction, and multiplication of fractions.

Problem-solving skills pupils might use:

   . Guess and check.

   . Eliminate possibilities.

Materials needed:

   . None

Comments and suggestions:

   . These 10 problems provide a great amount of computation practice with fractions. Be certain that pupils don't jump to conclusions too quickly regarding the smallest answer.

   . A few pupils may decide to "go in the hole" to get a smaller answer. For example, in problem 1, $\frac{2}{3} - \frac{4}{1} = -\frac{10}{3}$. Pupils with this kind of insight need additional praise and perhaps some of their results could be shared with the class.

Answers:

1. $\frac{1}{3} + \frac{2}{4} = \frac{5}{6}$

2. $\frac{2}{4} - \frac{1}{3} = \frac{1}{6}$

3. $\frac{1}{3} \times \frac{2}{4} = \frac{1}{6}$ or $\frac{1}{4} \times \frac{2}{3} = \frac{1}{6}$

4. $\frac{1}{4} + \frac{2}{8} = \frac{1}{2}$

Smallest Answer (cont.)

5. $\dfrac{\square}{\square} - \dfrac{\square}{\square} =$ Use
1, 2, 4, 8

6. $\dfrac{\square}{\square} \times \dfrac{\square}{\square} =$ Use
1, 2, 4, 8

7. $\dfrac{\square}{\square} + \dfrac{\square}{\square} =$ Use
2, 3, 4, 12

8. $\dfrac{\square}{\square} - \dfrac{\square}{\square} =$ Use
2, 3, 4, 12

9. $\dfrac{\square}{\square} \times \dfrac{\square}{\square} =$ Use
2, 3, 4, 12

10. $\dfrac{\square}{\square} + \dfrac{\square}{\square} + \dfrac{\square}{\square} =$ Use
1, 2, 3, 4, 6, 12

Answers: (cont.)

5. $\frac{4}{8} - \frac{1}{2} = 0$ or $\frac{8}{4} - \frac{2}{1} = 0$ or $\frac{2}{8} - \frac{1}{4} = 0$ or $\frac{1}{4} - \frac{2}{8} = 0$

6. $\frac{1}{4} \times \frac{2}{8} = \frac{1}{16}$ or $\frac{1}{8} \times \frac{2}{4} = \frac{1}{16}$

7. $\frac{2}{4} + \frac{3}{12} = \frac{3}{4}$

8. $\frac{2}{4} - \frac{3}{12} = \frac{1}{4}$

9. $\frac{2}{4} \times \frac{3}{12} = \frac{1}{8}$ or $\frac{2}{12} \times \frac{3}{4} = \frac{1}{8}$

*10. $\frac{1}{4} + \frac{2}{6} + \frac{3}{12} = \frac{5}{6}$

CLOSER TO

John and Sue were discussing their estimates for $12 \div 2\frac{5}{8}$. John estimated the answer to be 6. Sue estimated the answer to be 4. Who had the better estimate?

The teacher said that $12 \div 2\frac{5}{8} = 4\frac{4}{7}$. Since Sue's estimate was closest, she was asked to explain her procedure. Sue wrote this on the chalkboard.

My procedure also works for multiplication.

Study Sue's procedure. Use it to find a good estimate for each of these problems.

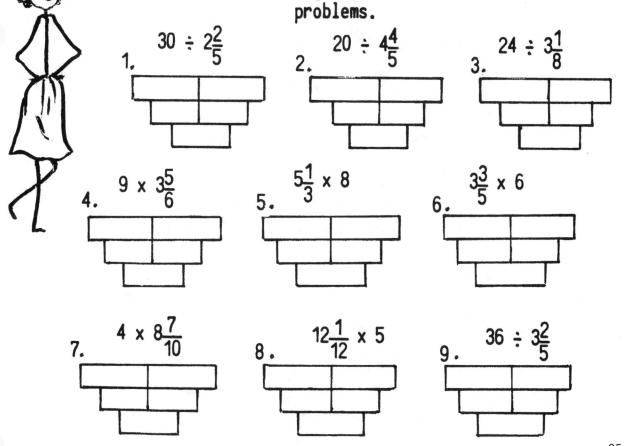

1. $30 \div 2\frac{2}{5}$

2. $20 \div 4\frac{4}{5}$

3. $24 \div 3\frac{1}{8}$

4. $9 \times 3\frac{5}{6}$

5. $5\frac{1}{3} \times 8$

6. $3\frac{3}{5} \times 6$

7. $4 \times 8\frac{7}{10}$

8. $12\frac{1}{12} \times 5$

9. $36 \div 3\frac{2}{5}$

<u>Closer To</u>

Mathematics teaching objectives:

. Estimate answers involving multiplication and division with fractions.

Problem-solving skills pupils <u>might</u> use:

. Clarify the problem through careful reading.
. Make reasonable estimates.
. Recognize limits.

Materials needed:

. None

Comments and suggestions:

. It may be necessary to go over the example with the entire class so they understand completely what to do.
. A culminating activity might be to have pupils work a similar list mentally (no pencils allowed!).
. When the activity is discussed, generalizations could be made explicit concerning the effects of increasing or decreasing either of the two numbers involved in the given exercises.
. You may wish to extend the discussion concerning estimation to whole numbers and decimals and to addition and subtraction.

Answers:

1. $30 \div 2\frac{2}{5}$ is closer to 15 than to 10.

2. $20 \div 4\frac{4}{5}$ is closer to 4 than to 5.

3. $24 \div 3\frac{1}{8}$ is closer to 8 than to 6.

4. $9 \times 3\frac{5}{6}$ is closer to 36 than to 27.

5. $5\frac{1}{3} \times 8$ is closer to 40 than to 48.

6. $3\frac{3}{5} \times 6$ is closer to 24 than to 18.

7. $4 \times 8\frac{7}{8}$ is closer to 36 than to 32.

8. $12\frac{1}{12} \times 5$ is closer to 60 than to 65.

9. $36 \div 3\frac{2}{5}$ is closer to 12 than to 9.

Grade 7

IV. DRILL AND PRACTICE - DECIMALS

Most seventh-grade classes are a collection of pupils with varying levels of skills. What can be done to provide additional practice and learning for all pupils? One solution is to offer drill and practice with decimals through problem-solving activities. While Ray is showing that .4 is less than .5, Jenny may be challenged to find as many decimals between .4 and .5 as she can.

One marvelous discovery in using problem-solving skills with a mixed class is that some pupils who are unskilled in computation are really good at seeing patterns or figuring out ways to solve problems. Some who hate ordinary drill problems will fill a page with computation to try to solve a problem.

Using The Activities

The activities in this section can be incorporated with the regular teaching and review of decimals. Most of them should be done in class so that the teacher can interact and give appropriate hints when necessary.

GOING IN CIRCLES

Rules

. Don't count the circles for "start" and "end."

. Always move to a larger number.

1. Find the longest path through the maze.

2. There are four different paths that are the shortest. Find all four of them.

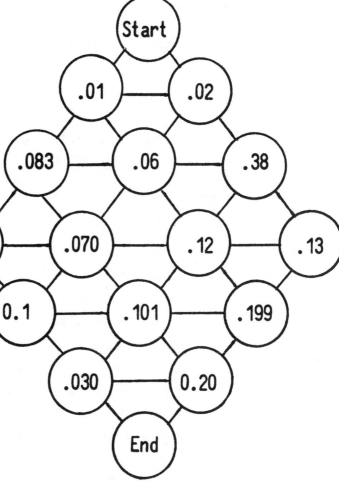

3. Find a path that is exactly six circles long.

4. Find a path that is exactly nine circles long.

5. Which circles will never be used? Explain.

6. Design your own maze. Give it to a classmate to work.

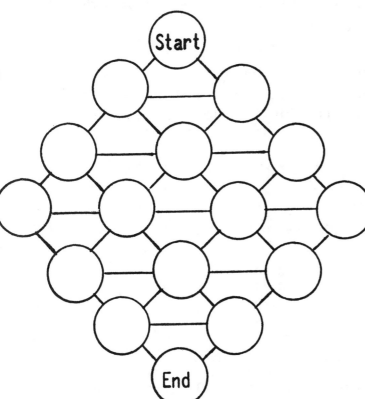

<u>Going</u> <u>In</u> <u>Circles</u>

Mathematics teaching objectives:

. Order decimals from the least to greatest,
. Determine when the placement of "0" does or does not affect the
  value of a decimal.

Problem-solving skills pupils <u>might</u> use:

. Guess and check.
. Eliminate possibilities.
. Create a problem situation similar to one already solved

Materials needed:

. None

Comments and suggestions:

. Until pupils carefully study the maze they probably will use a guess
  and check procedure. A pupil who has studied the puzzle carefully
  might recognize that the .38 and .030 circles are deadends, leaving
  12 non-deadend circles. This implies that if it is possible to go
  through these 12 circles this path will be the longest path.
. Encourage pupils to work individually.
. Encourage pupils to find other solutions.
. Some possible discussion questions could be:

    . How do we know that the shortest path must be 5 circles long?
    . Which circles should be avoided?   Why?
    . How do we know that the longest path cannot be greater than
      12 circles long?
    . How do we know that there is only one longest path?

Answers:

1. .01, .02, .06, .070, .083, .09, 0.1, .101, .12, .13, .199, 0.20

2. Four possible paths:
   .01,   .06,   .070,   .101,   0.20
   .02,   .06,   .12,   .199,   0.20
   .02,   .06,   .070,   .101,   0.20
   .01,   .06,   .12,   .199,   0.20

3. One possibility is:
   .01,   .06,   .070,   0.1,   .101,   0.20

4. .01, .06, .070, .083, .09, 0.1, .101, .199, 0.20

5. .38 and .030 (All numbers leading away from .38 are smaller;
                     all numbers leading in to .030 are larger.)

6. Answers will vary. Some of the more challenging ones could be
   given to the entire class.

GOING CRAZY WITH NUMBERS

Directions: Find the sum of <u>all</u> numbers in the flower. Before you begin, write about the method you plan to use.

<u>Going</u> <u>Crazy</u> <u>With</u> <u>Numbers</u>

Mathematics teaching objectives:

. Practice adding decimals.

. Practice mental arithmetic skills.

Problem-solving skills pupils <u>might</u> use:

. Break a complicated problem into manageable parts.

. Make and use a systematic listing.

. Solve a problem by using different procedures.

Materials needed:

. None

Comments and suggestions:

. Have pupils decide on a strategy to use. The importance of the first two specific skills listed below should be emphasized.

. Encourage pupils to check their answer by using a different procedure.

. Some will add each petal, leaf, and stem separately, record the partial sums and then find the grand total. Some may find each partial sum by tallying combinations of decimals which make one. Some may count all the .1, then the .2, etc.

Answer:

Starting with the petal in the 12 o'clock position and moving clockwise, the ten petals have individual sums of 3.9, 4.2, 5.5, 6.6, 5.1, 6.0, 5.7, 4.6, 6.0, and 5.5. The center of the flower is 7.0; the stem is 6.2; the left leaf is 3.4 and the right leaf is 3.9.

The <u>total</u> for the flower is 73.6.

A DECIMAL WEB

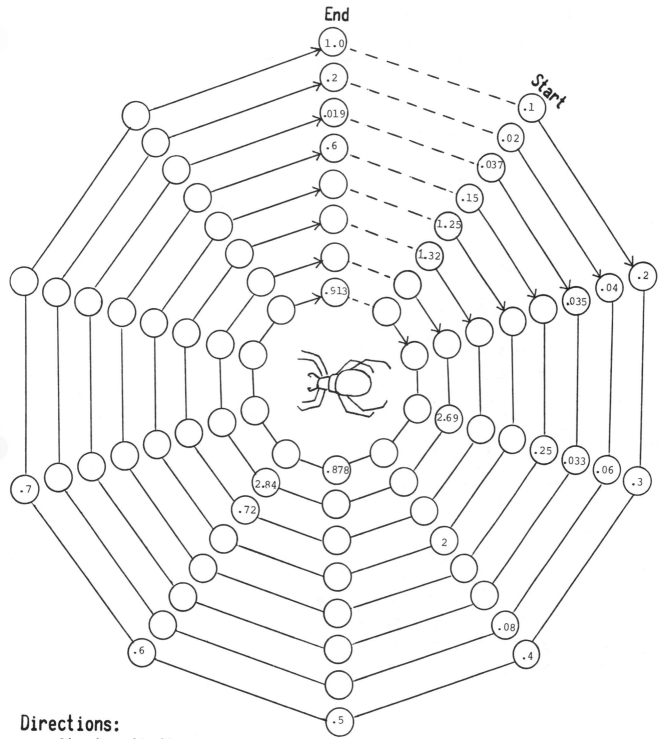

End

Start

Directions:
  . Start with the
    outermost part of the web.
  . Fill in the circles by moving in the direction of the arrows.
    Notice that most of the outer ring has been completed for you.
  . Each jump from one circle to the next must be the same. (In the
    outer ring the jump is .1.)   . Complete the rest of the web.

A Decimal Web
=============

Mathematics teaching objectives:
- Order and compute with decimals.
- Informally work with arithmetic series.

Problem-solving skills pupils might use:
- Identify patterns.
- Work backwards.

Materials needed:
- None

Comments and suggestions:
- You may need to clarify the directions for some pupils.
- The three outermost decagons will cause little difficulty. The other decagons will require a more sophisticated strategy. Pupils will need to determine the interval between successive terms. This can be done by finding the difference between two known numbers and then dividing by the number of intervals.

Answers:

(Numbers in outermost ring given first.)

.1,   .2,   .3,   .4,   .5,   .6,   .7,   .8,   .9,   1.0

.02,   .04,   .06,   .08,   .10,   .12,   .14,   .16,   .18,   .20

.037,   .035,   .033,   .031,   .029,   .027,   .025,   .023,   .021,   .019

.15,   .20,   .25,   .30,   .35,   .40,   .45,   .50,   .55,   .60

1.32,   1.20,   1.08,   .96,   .84,   .72,   .60,   .48,   .36,   .24

2.59,   2.64,   2.69,   2.74,   2.79,   2.84,   2.89,   2.94,   2.99,   3.04

1.25,   1.50,   1.75,   2.00,   2.25,   2.50,   2.75,   3.00,   3.25,   3.50

.850,   .857,   .864,   .871,   .878,   .885,   .892,   .899,   .906,   .913

MULTIPLYING FOR POINTS

1.8	4.7	.33	12.6	3.0
.9	.68	5.4	.40	2.03

Beth chose two different numbers from the above list and multiplied them. She then found the correct box below for her answer. The answer is worth the number of points indicated.

Example:
.9 x 4.7 = 4.23

Beth's answer goes in Box C because 4.23 is between 1.0 and 5.0. This problem is worth 3 points.

1. Beth found six multiplication problems whose answers are worth 1 point each. How many can you find? (You are not allowed to multiply a number by itself.)

2. Find as many multiplication problems as you can whose answers each are worth 4 points.

3. Find a multiplication problem whose answer fits

   a. in Box B.   c. in Box E.
   b. in Box C.   d. in Box F.

1 point	2 pts.	3 pts.	4 pts.	3 pts.	2 pts.	1 point
A	B	C	D	E	F	G
0 to .5	.5 to 1.0	1.0 to 5.0	5.0 to 10.0	10.0 to 20.0	20.0 to 40.0	40.0 to 70.0

Multiplying For Points
=====

Mathematics teaching objectives:

 . Multiply with decimals.

 . Practice mental arithmetic skills.

Problem-solving skills pupils <u>might</u> use:

 . Make reasonable estimates as answers.

 . Sort out and eliminate possibilities.

Materials needed:

 . None

Comments and suggestions:

 . Pupils may start multiplying two numbers at random. As they continue with the exercises this procedure is likely to lead to confusion and some frustration. At this time you might suggest using a rounding and estimating strategy.

 . This activity could be adapted to serve as a contest between teams of players. For example, one player selects two numbers and estimates in which box the answer will be. Then a calculator is used to see if the estimate was correct.

Answers:

1. There are seven problems worth 1 point each.

 $.33 \times .9 = .297$

 $.33 \times .68 = .2244$

 $.33 \times .4 = .132$

 $.9 \times .4 = .36$

 $.68 \times .4 = .272$

 $4.7 \times 12.6 = 59.22$

 $12.6 \times 5.4 = 68.04$

2. There are seven problems worth 4 points each.

 $1.8 \times 4.7 = 8.46$

 $1.8 \times 3 = 5.4$

 $1.8 \times 5.4 = 9.72$

 $4.7 \times 2.03 = 9.541$

 $12.6 \times .68 = 8.568$

 $12.6 \times .4 = 5.04$

 $3.0 \times 2.03 = 6.09$

3. One possible answer is given for each part.

 a. Box B $.33 \times 1.8 = .594$

 b. Box C $.9 \times 4.7 = 4.23$

 c. Box E $.9 \times 12.6 = 11.34$

 d. Box F $12.6 \times 2.03 = 25.578$

HONE ON THE RANGE

Get: One calculator for every two players

Rules:

. This activity is best for two players.

. Decide on a range, say between 730 and 760 or a smaller range like 600 to 605. The object is to get the answer to a multiplication problem to be within the range.

. Player A enters some number (a two-digit number might be best) on the calculator and pushes the X key.

. Player B enters a number and pushes the X key, trying to get the answer to be within the range.

. Suppose player B's answer is not within the range. Then player A enters a number and pushes the X key, trying to get the answer to be within the range.

. Play continues until one of the players does get an answer within the range.

Other ranges to try: 850-855,   175-180,   335-340

Superhard: 199-200,   3000-3001

EXTENSION

Do the same activity but this time use the ÷ key.

Pick a low range, like 40-45.
Start with a large first number, like 638.

<u>Hone</u> <u>On</u> <u>The</u> <u>Range</u>

Mathematics teaching objectives:

. Practice mental arithmetic skills using decimals.

. Round decimals.

. Investigate multiplication by decimals greater than or less than 1.

Problem-solving skills pupils <u>might</u> use:

. Make reasonable estimates.

. Recognize limits and/or eliminate possibilities.

. Guess and check.

Materials needed:

. One calculator for every two players.

Comments and suggestions:

. You may need to go through an example with the entire class before they begin working on their own.

. This is an excellent example of the use of a calculator to promote significant instructional objectives.

. If enough calculators are available, this could be an activity done individually. Pupils could see how many steps it took them to "hone in on the range."

. Since the calculator is used, pupils probably will enjoy playing the game on different occasions.

Grade 7

V. PERCENT SENSE

V.   PERCENT SENSE

How would you expect junior high pupils to solve this problem?

Every bike is 15% off the regular price.

REG. PRICE $200

Some pupils might attempt a proportion method but may have forgotten which numbers go on top. Others might change the percent to a decimal and multiply (or maybe even divide). Some may try translating the problem into an equation and, no doubt, several will say, in frustration, that they never really understood percents.

Hopefully, upon completion of this unit, pupils <u>will</u> have a good feeling for percents. They will have developed a kind of "percent sense," something that too often has been lacking in their experience.

In the bicycle problem, as with many percent problems, pupils should be encouraged to use mental computations. (Some authorities say that 75% of the adult non-occupational uses of mathematics involve mental arithmetic and estimations.)

. Here is one "percent sense" method that could be used on the bicycle problem:

15% off means $15 off of every $100, or $30 off of $200.

Thus, the selling price is $170.

It is not the intention of this unit to develop one particular method for solving percent problems. Instead, pupils will have many intuitive experiences which involve using charts, ratio, proportion, the "per-hundred" method, and the "one-percent" method. A more formal development of any of these procedures should be postponed until the entire unit is completed.

Three different models are used to develop percent concepts:

. The 100-grid---probably the easiest model for working with percents less than 100.

. The circle model---a good one for developing the 100% quantity.
. The number line---effective when considering quantities over 100%.

Using The Activities

The development in this unit relies only on intuitive understanding of fractions and decimals. Pupils will, however, need some previous experience in multiplying and dividing by powers of ten, especially 100.

Some lessons have "class exercises." These are intended to be completed and discussed during class time. The exercises could be assigned as homework. Some lessons are short and could easily be combined with others to make a full period activity.

The sequence of lessons in this unit is important. The lessons should be presented in the order given.

SENSE OR NONSENSE

You already know a lot about percent. Decide whether the following statements are reasonable. Explain why.

1. Mr. Bragg says he is right 100% of the time.
   Do you think Mr. Bragg is bragging? Why?

2. The Todd family ate out last Saturday. The bill was $24.
   Mr. Todd decided to leave a 50% tip.
   Do you think it is too much? Why?

3. Joe loaned Jeff a dollar. He said the interest would be 75% a day. Is this a pretty good deal for Joe? Why?

4. Cindy spends 100% of her allowance on candy.
   Do you think this is sensible? Why?

5. The "Never Miss" basketball team made 10% of the baskets they tried. Do you think they should change their name? Why?

6. Sarah missed 10 problems on the science test. Do you think her percent is high enough for her to earn an A? Why?

7. Billie has a paper route. She gets to keep 25% of whatever she collects. Do you think this is a good deal? Why?

8. The weatherman said, "There's a 100% chance of rain for tomorrow." Do you think this weatherman lives in Oregon? Why?

9. Miss Green was complaining, "Prices have gone up at least 200% this past year." Do you think she is exaggerating? Why?

10. Write down two more percent problems like those above.
    Give them to some classmates to get their reactions.

Sense Or Nonsense

Mathematics teaching objectives:

. Address a variety of questions relating to familiar percent situations.

Problem-solving skills pupils might use:

. Make decisions based upon data.

. Make reasonable estimates as answers.

. Make explanations based upon conditions of the problem.

Materials needed:

. None

Comments and suggestions:

. Pupils should be able to answer the questions in this introductory lesson by relying upon the knowledge they've acquired from their frequent informal experiences with percent situations.

. One objective of the lesson is to generate some interesting class discussion regarding percent. Thus, it is best used in a class discussion setting.

. The lesson is not intended to take a full period.

. Many pupils will skip question 10 unless they expect special attention will be given to their answers.

Answers: Answers will vary. Some possible answers are given.

1. Yes. Nobody's perfect.
2. Probably. The usual tip is 10% to 15%.
3. Yes. This would amount to 75¢ per day.
4. It depends upon how much her allowance is.
5. Probably. A 10% shooting record isn't very good.
6. It depends on how many problems were on the test. For example, an "A" grade for 10 wrong out of 20 problems would be unusual.
7. Probably. Most paper carriers get quite a bit less.
8. Maybe. It does rain a lot in Oregon.
9. Yes. Although prices are going up, a 200% increase is excessive. This means an item that sold for $1.00 now would be selling for $3.00.
10. Answers will vary.

ROOM DESIGNS

1. Mr. Owen designed a new science area for the school. Notice that it is divided up into 100 parts. The office takes up 9% of the total space.

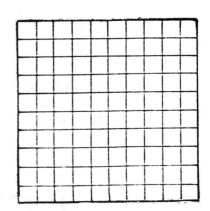

   a. What percent of the space is occupied by each of the other rooms?

   b. Mr. Owen decided to put a closet inside the lecture room. Show where it might be placed. Make it a "3% closet."

   c. Four separate lab tables take up 6% of the total space of the entire science area. Carefully place these four tables in the laboratory section.

2. This 100-grid represents the basement of a home.

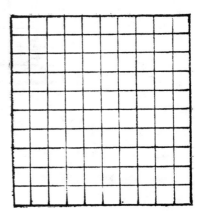

   a. Show a good way to divide up the space so that

      . 15% is occupied by a bathroom.
      . 20% is occupied by a darkroom.
      . 21% is occupied by a storeroom.
      . the remaining percent is occupied by a recreation room.

   b. What percent of the basement is your recreation room?

3. This 100-grid represents the first floor of an office building. Design a floor plan so that

      . 68% of the space is for 4 separate offices.
      . 20% of the space is for hallways.
      . 12% of the space is for 2 separate bathrooms.

Room Designs

Mathematics teaching objectives:

. Solve percent problems using a 100-grid.

Problem-solving skills pupils might use:
. Use a drawing.
. Create a design with certain specifications.

Materials needed:

. None

Comments and suggestions:

. Emphasize the need for careful planning in Exercises 2 and 3. For example, long, narrow rooms should be avoided. Bathrooms and hallways need to have reasonable locations.

. This is not intended to be a full-period lesson.

Answers:

1. a. Storeroom - 12%
        Lecture Room - 42%
        Laboratory - 37%

    b. Answers will vary.

    c. Answers will vary. The lab tables all do not have to be the same size. If they are, each has to be $1\frac{1}{2}$%   ($1\frac{1}{2}$ squares).

2. a. This is one possible arrangement.   3. One possible room arrangement is given.

    b. 44%

Possible Extension:

Consider adding a 20% room. The new plan will be 120% of the original.

USING A PERCENT FINDER

Estimate the percent of the square that is shaded. Then use a transparent 100-grid to check your estimate. Record both results. How close was your estimate?

☺ Good   ☺ Fair   ☹ Awful

1.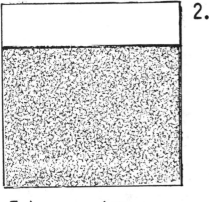

Est.____ Ans.____

2.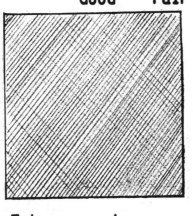

Est.____ Ans.____

3.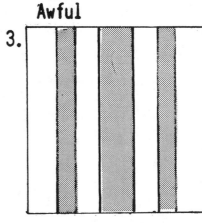

Est.____ Ans.____

4.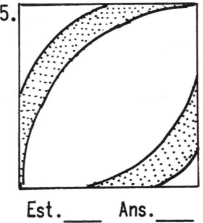

Est.____ Ans.____

5.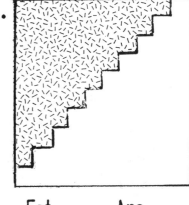

Est.____ Ans.____

6.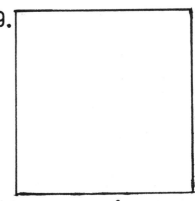

Est.____ Ans.____

7.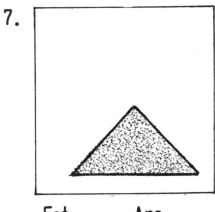

Est.____ Ans.____

8.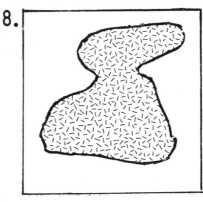

Est.____ Ans.____

9.

Make up one of your own. Give it to a classmate to figure out.

Using A Percent Finder

Mathematics teaching objectives:
.   Make estimates involving percent.
.   Determine percents using a 100-grid.

Problem-solving skills pupils might use:
.   Use a drawing and physical model.
.   Break a problem into parts or steps.
.   Make reasonable estimates as answers.

Materials needed:
.   A transparent 100-grid for each pupil.   See page 4 of the pupil materials.

Comments and suggestions:
.   Pupils need to make estimates before the grids are passed out.   They
    should recognize that the estimates they make are not wrong.   Some,
    however, may be better than others.

.   Pupils will discover different ways to estimate when using the percent
    finder.   For example, some will combine half-squares; others will count
    only squares that are half or more.   The various methods should be dis-
    cussed.

.   This activity is short and could be used easily in conjunction with
    the following lesson.

Answers:
1.   80%
2.   100%
3.   40%
4.   88%   (One method is to subtract the unshaded squares from 100.)
5.   27%
6.   45%
7.   12%
8.   40%
9.   Answers will vary.

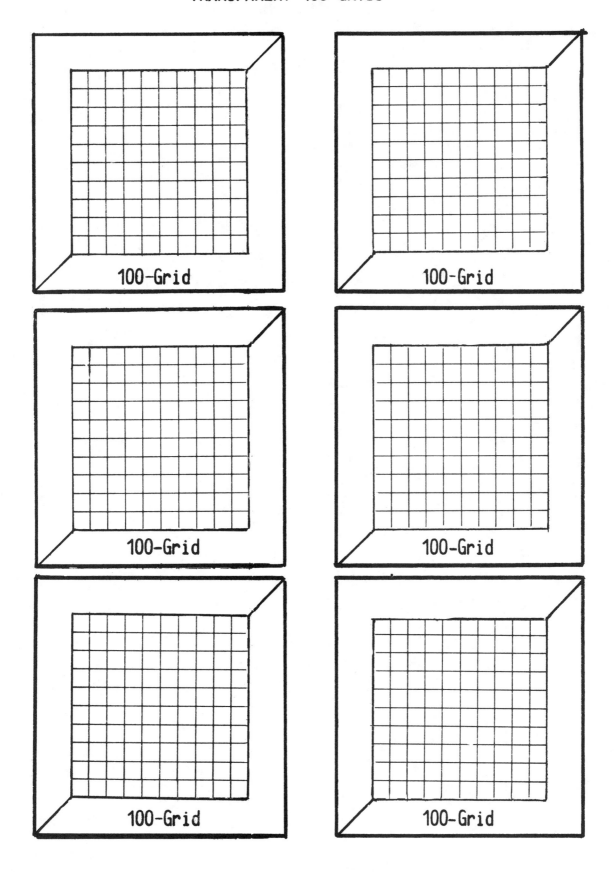

100-Grid

100-Grid

100-Grid

100-Grid

100-Grid

100-Grid

USING A 100-GRID

How much of the 100-grid does each shaded figure show? First make an estimate. Then use a transparent 100-grid to check your estimates. Record both results.

This is a 100-grid.

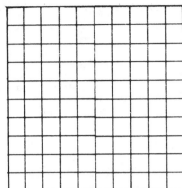

1.

Est.____

Ans.____

2.

Est.____   Ans.____

3.

Est.____   Ans.____

4.

Est.____

Ans.____

5.

Est.____   Ans.____

6.

Est.____

Ans.____

7.

Est.____ Ans.____

8.

Est.____   Ans.____

9.

Est.____   Ans.____

dea from Ratio, Proportion and Scaling, Creative Publications
© PSM 81

Using A 100-Grid

Mathematics teaching objectives:
- . Make estimates involving percent.
- . Determine percents using a 100-grid.

Problem-solving skills pupils _might_ use:
- . Use a drawing and physical model.
- . Break a problem into parts or steps.
- . Make reasonable estimates as answers.

Materials needed:
- . A transparent 100-grid for each pupil. See page 4 of the pupil materials.

Comments and suggestions:
- . Pupils need to make estimates before the grids are passed out.

- . This lesson is short and could be used easily with the previous lesson.

Answers:

1. 13%
2. 18%
3. 36%
4. 34%
5. 40.5%
6. 32%
7. 10%
8. 56%
9. 156%   (Perhaps pupils could draw other regions that are over 100%
        and estimate and then determine the actual percent.)

100 PERCENT

Sammy read this article in the daily <u>Bayshore</u> <u>Bugle</u>.

Do you think there is a mistake in this article? Why?

> 35% of the people in Bayshore are under 25 years of age.
> 45% are between 25 and 40.
> 40% are over 40.

1. Make up two of your own newspaper articles which have percents in them. The first one should make sense. The second one should not.

2. Jane was coloring some squares. She needed to color them according to the percents in the chart.

   In some cases, she had a choice as to how they should be colored.

   See if you can help Jane complete the chart.

Square	Blue	Green	Brown	Not Colored
1st	10%	25%	15%	
2nd	31%	7%		10%
3rd	19%		36%	45%
4th		8%	12%	61%
5th	22%			53%
6th		26%	41%	
7th	40%	36%	35%	

3. Do these circle graphs seem reasonable? Explain your answer.

a.
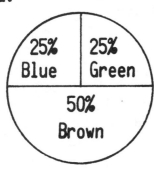

b.
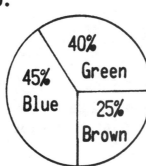

c.
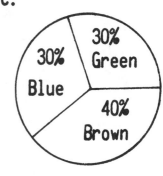

100 Percent

Mathematics teaching objectives:

. Detect situations in which percent is incorrectly used.

. Estimate part of a unit using percent terminology.

. Recognize that 100% is the whole thing.

Problem-solving skills pupils *might* use:

. Use a chart.

. Make reasonable estimates as answers.

. Detect errors.

. Work backwards.

Materials needed:

. None

Comments and suggestions:

. In the last few lessons, a 100-grid was used to give pupils an intuitive understanding of percent. In this lesson, a circle model is used. This model is useful when pupils are dealing with 100% quantities.

. To check the estimates in Exercises 4 and 5, you might provide some "circle transparencies" divided into ten equal parts to lay over the estimates.

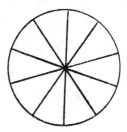

Answers:

1. Answers will vary.

2. 1st - 50% not colored

   2nd - 52% brown

   3rd - 0% green

   4th - 19% blue

   5th - The sum of <u>green</u> and <u>brown</u> must be 25%.

   6th - The sum of <u>blue</u> and <u>not</u> <u>colored</u> must be 33%.

   7th - Not possible. The total is already more than 100%.

(Answers continued on next page...)

4. Guess what percent of each circle is labeled red, yellow, and orange.

a.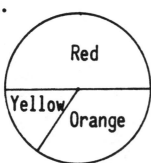

b.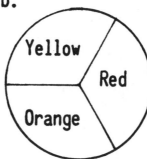

c.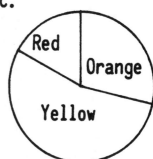

5. Draw lines which approximately divide each circle according to the percents given.

a.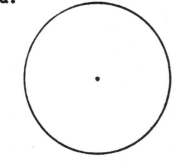

Red	50%
Yellow	40%
Orange	10%

b.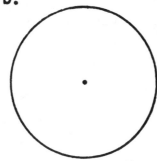

Red	25%
Yellow	3%
Orange	72%

c.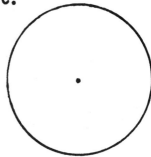

Red	85%
Yellow	10%
Orange	5%

d.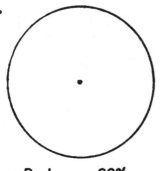

Red	60%
Yellow	20%
Orange	20%

e.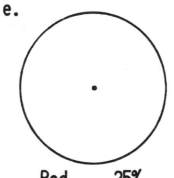

Red	25%
Yellow	25%
Orange	25%
Brown	25%

f.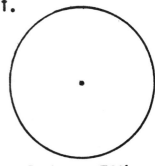

Red	50%
Yellow	25%
Orange	20%
Brown	5%

Answers: (cont.)

3. a. Reasonable. The sum is 100% .

   b. Not reasonable. The sum is more than 100% .

   c. Reasonable. The sum is 100%.

4. Answers are approximate. Some possible answers are given.

   a. Red - 50%   b. Red - $33\frac{1}{3}$%   c. Red - 17%

      Yellow - 15%   Yellow - $33\frac{1}{3}$%   Yellow - 54%

      Orange - 35%   Orange - $33\frac{1}{3}$%   Orange - 29%

5. a. 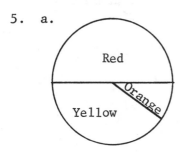   b. 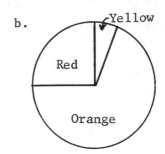   c.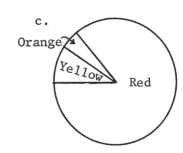

   d. 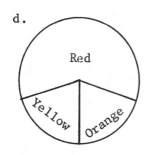   e. 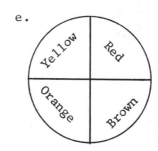   f.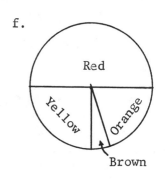

PERCENTS ON A NUMBER LINE

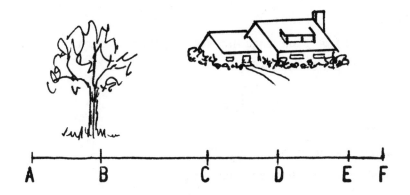

Joe has been hired to help with a ditch-digging project. A certain percent of the total ditch will be completed at the end of the day. AF is the whole ditch, or 100%.

Class Exercises

1. Make some estimates (in percents). Record your estimates in the table. Remember, in each case your estimate is from point A.

		Estimate
Monday	A to B	%
Tuesday	A to C	%
Wednesday	A to D	%
Thursday	A to E	%
Friday	A to F	%

2. Now use this scale to find the exact percents. Place this information in the table next to your estimates. How close were were your estimates?

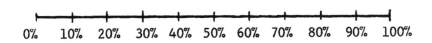

Percents On A Number Line

Mathematics teaching objectives:

. Estimate parts of a line segment using percent terminology.

. Use percents greater than 100.

Problem-solving skills pupils might use:

. Use a diagram.

. Make decisions based upon data.

. Make reasonable estimates as answers.

. Make necessary measurements for checking a solution.

Materials needed:

. Metric rulers could be provided for pupils to check their estimates.

Comments and suggestions:

. This is the first time pupils have encountered the number-line model in working with percent. Pupils should be aware that the 100% quantity can change from problem to problem.

. The number-line model also makes it possible to discuss, with meaning, quantities over 100%.

. An easy way to check the pupils' work is to make a transparency to lay over the answers.

Answers:

Class Exercises   (Answers are approximate)

1. and 2.   Monday   - 20%

Tuesday   - 50%

Wednesday - 70%

Thursday   - 90%

Friday   - 100%

Percents On A Number Line (cont.)

3. Each ditch is a different length. Estimate the percents that go with each letter. Write your estimate below the letter.

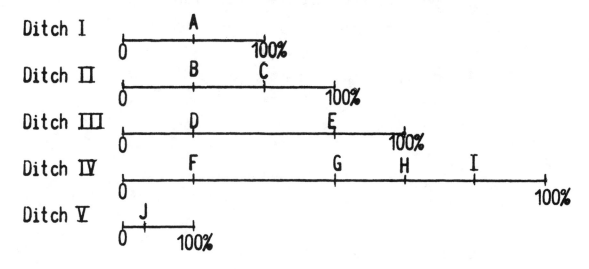

Exercises

1. On <u>each</u> segment, estimate and label the

| 50% point
with an A | 25% point
with a B | 75% point
with a C |

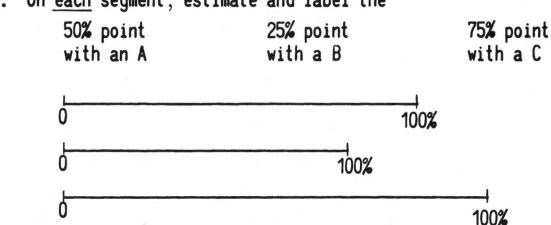

2. Estimate and label the 100% point on each segment.

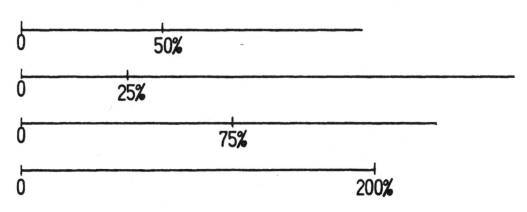

<u>Percents</u> <u>On</u> <u>A</u> <u>Number</u> <u>Line</u>

Answers: (cont.)

<u>Class</u> <u>Exercises</u> (Answers are approximate)

3.   A - 50%   F - 17%

      B - 33%   G - 50%

      C - 67%   H - 67%

      D - 25%   I - 83%

      E - 75%   J - 35%

<u>Exercises</u>   (Answers are approximate)

1.

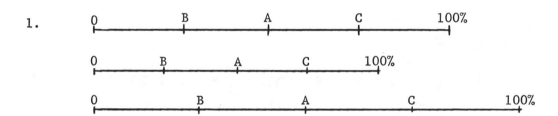

2.

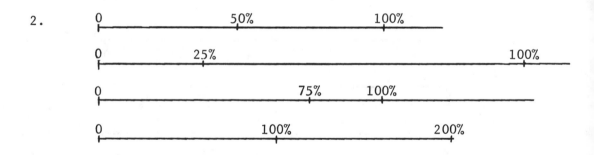

Percents On A Number Line (cont.)

3. Estimate and label the 150% point on each segment.

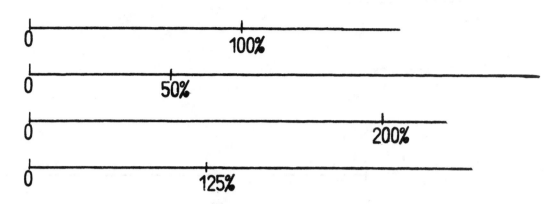

4. On the ditch below, estimate and label

    a. the 50% point with an A.

    b. the 25% point with a B.

    c. the 75% point with a C.

5. Suppose the ditch in Exercise 4 is 400 feet long.
    On the same ditch, estimate and label

    . the 200-foot point.
    . the 100-foot point.
    . the 50-foot point.
    . the 300-foot point.

Answers: (cont.)

Exercises

3.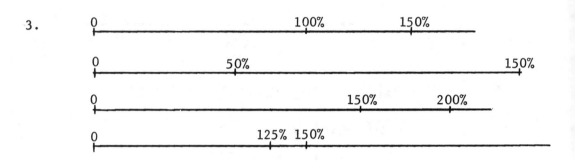

4.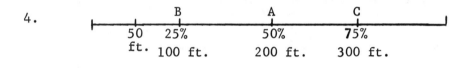

5. See Exercise 4.

USING CHARTS

Class Exercises

1. Big Willie likes 60¢ hamburgers. Complete the chart to show various costs of hamburgers for Big Willie.

Number of hamburgers	1	2	4	8	16	20	40
Cost (in cents)	60						

Use your completed chart to determine the cost of

a. 3 hamburgers.   c. 12 hamburgers.
b. 6 hamburgers.   d. 60 hamburgers.

2. Slim works in a donut shop. The donuts sell for $2.40 per dozen. Slim decided to make a chart to show how much different amounts of donuts would cost. Show how to complete the chart below.

Donuts	12	11	10	9	8	7	6	5	4	3	2	1
Cost (in cents)	240											

Use your completed chart to help you determine the cost of

a. 24 donuts.   b. 18 donuts.   c. 21 donuts.   d. 40 donuts.

Exercises

1. Pencils are selling for 4 for 88¢. Complete the chart.

Pencils	1	2	3	4	8	16	32
Cost							

2. Bananas are 3 pounds for 99¢. How much will 5 pounds cost?

Using Charts

Mathematics teaching objectives:

. Solve problems by using the concept of ratio (rate).

. Solve story problems and record answers systematically in a table.

Problem-solving skills pupils might use:

. Make and use a systematic listing in a table.

. Identify trends suggested by data in a table.

. Make predictions based upon patterns.

Materials needed:

. None

Comments and suggestions:

. The use of charts provides readiness for the ratio-proportion method often used to solve percent problems. No formal ratio-proportion work is expected at this time. In fact, it is recommended that any formal development be postponed until the entire unit is completed.

. Sometimes a picture or diagram is helpful to illustrate rate pairs:

e.g.   60¢   } 120¢

. This lesson provides many opportunities for using interpolation and extrapolation.

Answers:

Class Exercises

1.

Hamburgers	1	2	4	8	16	20	40
Cost	60	120	240	480	960	1200	2400

    a. 180   b. 360   c. 720   d. 3600

Note: Encourage pupils to use the charts to find these answers rather than multiplying. There are many ways to solve Exercises a-d. For example, for (c) you could add the costs of 4 hamburgers and 8 hamburgers - or you could subtract the cost of 4 hamburgers from the cost of 16 hamburgers. These methods are applications of the distributive property and should be discussed.

(Answers continued on next page...)

3. Candy bars are on sale for 3 for 70¢. Use a chart to show the cost of

   a. 6 candy bars.

   b. 9 candy bars.

   c. 7 candy bars.

4. Sandy plays basketball for the school team. She leads the league in free-throw accuracy. So far Sandy has made 14 out of 20. Suppose she continues at this rate. Complete the chart to show what her record would look like.

Free-throws made	14				
Free-throws tried	20	40	50	60	100

Answers:

Class Exercises (cont.)

2.

Donuts	12	11	10	9	8	7	6	5	4	3	2	1
Cost	240	220	200	180	160	140	120	100	80	60	40	20

a. 480   b. 360   c. 420   d. 800

Note: Encourage pupils to use their completed charts. The various methods used to solve Exercises a-d should be discussed.

Exercises

1.

Pencils	1	2	3	4	8	16	32
Cost	22	44	66	88	176	352	704

2. $1.65

3.

Number of Candy Bars	3	6	9	12
Cost In Cents	70	140	210	280

4.

Free Throws Made	14	28	35	42	70
Free Throws Tried	20	40	50	60	100

PER HUNDRED

In a recent baseball series -

Homer Wrun had 7 hits out of 20 times at bat.
Grant Schlamme had 9 hits out of 25 times at bat.

Make a guess as to who had the best record.

Class Exercises

1. Assume that they continue to hit at the same rate.

   a. Complete the two charts for Homer and Grant.

   Homer

Hits	7				
Times at bat	20	40	60	80	100

   Grant

Hits	9			
Times at bat	25	50	75	100

   b. According to the charts -

   Homer would get ___ hits per 100 times at bat.
      What percent is this? _____

   Grant would get ___ hits per 100 times at bat.
      What percent is this?

   c. Who had the best record?

2. Lefty Fielder had 8 hits in 40 times at bat.

   a. At this rate, how many hits would he have in 100 times
       at bat? Complete the chart to find out.

Hits	8				
Times at bat	40	20	10	50	100

   b. What percent of the time did Lefty get a hit?

Per Hundred

Mathematics teaching objectives:

    . Determine percent by using tables.

    . Solve percent problems by using concept of ratio (rate).

Problem-solving skills pupils _might_ use:

    . Make a systematic listing in a table.

    . Identify trends suggested by data in tables.

    . Draw conclusions based upon observed patterns.

Materials needed:

    . None

Comments and suggestions:

    . As in the previous lesson, the use of charts provides readiness for
the ratio-proportion method often used in solving percent problems.
The main emphasis here is in building up and/or breaking down entries
in a chart to eventually read the "per hundred" numbers.

Answers:

Class Exercises

1. a.

Homer

Hits	7	14	21	28	35
Times At Bat	20	40	60	80	100

Grant

Hits	9	18	27	36
Times At Bat	25	50	75	100

   b. Homer, 35 hits - 35% ;   Grant, 36 hits - 36%

   c. Grant

2. a.

Hits	8	4	2	10	20
Times At Bat	40	20	10	50	100

   b. 20%

(Answers continued on next page...)

Per Hundred (cont.)

Exercises

1. Determine the percent of each of the following. Complete
   the chart for each one.

   a. 8 hits in 20 times
       at bat _____ %

Hits	8				
Times at bat	20	40	60	80	100

   b. 4 hits in 16 times
       at bat _____ %

Hits	4			
Times at bat	16	8	4	100

   c. 2 hits in 5 times
       at bat _____ %

Hits	2			
Times at bat	5	10	50	100

   d. 6 hits in 30 times
       at bat _____ %

Hits	6			
Times at bat	30	10	50	100

   e. 24 hits in 80 times
       at bat _____ %

H	24				
AB	80	40	20	10	100

   f. 60 hits in 150 times
       at bat _____ %

H	60		
AB	150	300	100

2. Make a chart. Find the percent.
   a. 21 problems correct out of 25 assigned

   b. 9 baskets made out of 15 tries

Answers: (cont.)

Exercises

1. a.

H	8	16	24	32	40
AB	20	40	60	80	100

40%

b.

H	4	2	1	25
AB	16	8	4	100

25%

c.

H	2	4	20	40
AB	5	10	50	100

40%

d.

H	6	2	10	20
AB	30	10	50	100

20%

e.

H	24	12	6	3	30
AB	80	40	20	10	100

30%

f.

H	60	120	40
AB	150	300	100

40%

2. Two possible charts are given.

a.

Problems Correct	21	42	84
Problems Assigned	25	50	100

84%

b.

Baskets Made	9	3	6	60
Baskets Tried	15	5	10	100

60%

3. One of many possibilities:

Player	Hits	Times at Bat	Percent
Sue	2	10	20
Jill	6	8	75
Joan	0	9	0
Jan	6	12	50
Pam	4	10	40
Kim	2	20	10
Bev	14	20	70
Beth	9	10	90
Nan	8	8	100

Per Hundred (cont.)

3. All of the nine players on a baseball team have different bat-
    ting records (percents). Fill in the chart below to show how this
    could be possible. Have a classmate check your completed
    chart.

Player	Hits	Times at Bat	Percent
Sue			
Jill			
Joan			
Jan			
Pam			
Kim			
Bev			
Beth			
Nan			

I'VE GOT YOUR NUMBER

Get a page from a telephone book.

1. Look at the last four digits of some of the numbers. What digit do you think is used most often? Make a guess.

2. Tally the last four digits of 50 telephone numbers. Use the chart below.

Digit	Tally	Total
0		
1		
2		
3		
4		
5		
6		
7		
8		
9		

3. What digit occurred most often?
   What percent is this?

4. What digit occurred least often?
   What percent is this?

5. Write a percent for each of the other digits.

6. Combine your results with your classmates'. Is there a digit that occurs most often?

7. Repeat the activity with a grocery ad. Tally prices until you have tallied 50 digits.

   a. What digit occurred most often? What percent is this?

   b. What digit occurred least often? What percent is this?

   c. Comment about your results.

<u>I've</u> <u>Got</u> <u>Your</u> <u>Number</u>

Mathematics teaching objectives:

. Determine percent from data collected in a laboratory activity.

. Solve percent problems by using concept of ratio (rate).

Problem-solving skills pupils <u>might</u> use:

. Collect data needed to solve problems.

. Record data in a table.

. Make necessary computations needed for the solution.

. Be aware of other solutions.

Materials needed:

. Pages from a telephone book

Comments and suggestions:

. This activity could be introduced by asking if there is a most popular digit if we consider the first three digits of a set of telephone numbers.

. Pupils should work in pairs with one person reading off numbers and the other keeping the tally.

Answers:

1. to 5.   Answers will vary. The table shows the results of one experiment.

Digit	Tally	Total	Percent
0	‖‖‖‖‖‖‖‖‖ I	21	10.5%
1	‖‖‖‖‖‖‖‖‖‖ I	26	13 %
2	‖‖‖‖‖‖‖‖‖‖ I	26	13 %
3	‖‖ I	11	5.5%
4	‖‖‖‖	20	10 %
5	‖‖‖‖ ‖‖	24	12 %
6	‖‖‖ ‖	18	9 %
7	‖‖‖‖ ‖‖	24	12 %
8	‖‖‖ ‖	17	8.5%
9	‖‖ ‖	13	6.5%

6.   Answers will vary.

7.   Probably <u>9</u> will be the most popular digit. To some persons, a 59¢ price may appear much less expensive than one priced at 60¢.

PERCENT SHORTCUTS

Jane was given lots of problems. All of them required that she find various percents of $200. Jane made a table to help her solve the problems.

What is 50% of $200?	$100
What is 25% of $200?	___
What is 10% of $200?	___
What is 5% of $200?	___
What is 1% of $200?	___

Class Exercises

1. Complete the table.

2. Use the table to help you solve these problems.
   a. 6% of $200
   b. 9% of $200
   c. 23% of $200
   d. 41% of $200
   e. 87% of $200

Exercises

1. Complete this table. Then use the table to help you solve the problems.

What is 50% of 500?	___
What is 25% of 500?	___
What is 10% of 500?	___
What is 5% of 500?	___
What is 1% of 500?	___

   a. 8% of 500
   b. 12% of 500
   c. 33% of 500
   d. 60% of 500
   e. 95% of 500

Make your own tables to help you solve the problems below.

2. a. 7% of 1600
   b. 11% of 1600
   c. 39% of 1600
   d. 67% of 1600
   e. 85% of 1600

3. a. 4% of $800
   b. 9% of $800
   c. 21% of $800
   d. 49% of $800
   e. 90% of $800

4. a. 3% of $600
   b. 24% of $600
   c. 40% of $600
   d. 65% of $600
   e. 99% of $600

5. a. 15% of $160
   b. 30% of $160
   c. 51% of $160
   d. 70% of $160
   e. 89% of 160

Percent Shortcuts

Mathematics teaching objectives:

. Use tables to find the percent of a number.

Problem-solving skills pupils _might_ use:

. Make and use a systematic listing in a table.
. Identify patterns suggested by data.
. Make necessary computations needed for the solutions.

Materials needed:

. None

Comments and suggestions:

. Discuss the various ways pupils used to complete the tables. For example, some pupils probably will think of 50% as $\frac{1}{2}$, 25% as $\frac{1}{4}$, 10% as $\frac{1}{10}$, etc.

. Discuss how the tables were used to solve the problems. For example, in Class Exercise 2c, pupils might add two 10%s and three 1%s, or they might subtract two 1%s from a 25%.

Answers:

Class Exercises   1.

What is 50% of $200?	$100
25%	200	($ 50)
10%	200	($ 20)
5%	200	($ 10)
1%	200	($ 2)

2. a. $12
   b. $18
   c. $46
   d. $82
   e. $174

Exercises

1.

What is 50% of 500?	(250)
25%	500	(125)
10%	500	(50)
5%	500	(25)
1%	500	(5)

a. 40
b. 60
c. 165
d. 300
e. 475

2. Possible Table:

50% of 1600 = 800
25%	1600	400
10%	1600	160
5%	1600	80
1%	1600	16

a. 112
b. 176
c. 624
d. 1072
e. 1360

4. a. $ 18
   b. $144
   c. $240
   d. $390
   e. $594

3. Possible Table:

50% of 800 = 400
25% of 800 = 200
10%	800	80
5%	800	40
1%	800	8

a. $ 32
b. $ 72
c. $168
d. $392
e. $720

5. a. $ 24
   b. $ 48
   c. $ 81.60
   d. $112
   e. $142.40

ONE-PERCENT METHOD

Class Exercises

1. Sometimes it's useful to know 1% of a number. Complete the
    1% chart below.

    1% of $100 is $ 1.00   1% of $500 is $____
    1% of $200 is $ 2.00   1% of $550 is $____
    1% of $300 is $____   1% of $570 is $____
    1% of $400 is $____   1% of $1200 is $____

2. Use your completed chart to help you solve this problem.

    Ann earned $200 last summer. She saved 9% of her
    earnings. How much did she save?

3. Complete this earnings chart. Use the 1% chart above to
    help you.

	Earnings	% Saved	Amount Saved
Ann	$ 200	9%	
Mary	$ 200	12%	
John	$ 300	8%	
Kay	$1200	6%	
Wayne	$ 500	11%	
Ken	$ 550	10%	
Wanda	$ 570	12%	
Roy	$ 400	21%	
Barbara	$ 800	7%	
Walt	$1000	4%	
Keith		5%	$ 20
Sue		6%	$ 30
Ronnie	$ 600		$ 18
Sophie	$ 700		$ 28

One-Percent Method

Mathematics teaching objectives:

. Use one-percent method to solve percent problems when presented in table or story problem form.

Problem-solving skills pupils might use:

. Break a problem into parts.
. Work backwards.
. Make computations necessary for solution.
. Create a problem situation which can be solved by a given procedure.

Materials needed:

. None

Comments and suggestions:

. Pupils use the "one-percent" method to solve the problems. Two prerequisites are necessary:
    a. an understanding that 1% means one-hundredth
    b. skill in finding one-hundredth of a number (dividing by 100).

. Many percent problems are easy to solve by the "one-percent" method and the method usually makes sense to pupils. For example, to find 12% of $600, you first find 1% of $600 and multiply your answer by 12. You'll no doubt need to give some direction as to how to apply this method, especially when dealing with story problems.

. This lesson and the following one can probably be done in one class period.

Answers:

Class Exercises

1. $1.00
    $2.00
    $3.00
    $4.00
    $5.00
    $5.50
    $5.70
    $12.00

2. $18

3.

	Earnings	% Saved	Amount Saved
Ann			$18
Mary			$24
John			$24
Kay			$72
Wayne			$55
Ken			$55
Wanda			$68.40
Roy			$84
Barbara			$56
Walt			$40
Keith	$400		
Sue	$500		
Ronnie		3%	
Sophie		4%	

Exercises

1. a. $128   b. $48   c. $108   d. $405   e. $72   f. $9

2. Answers will vary.

One-Percent Method (cont.)

<u>Exercises</u>

1. Solve these problems. Use the one-percent method.

   a. Mr. Rich earned $1600 one month. He spent 8% on food.
       How much did he spend on food?

   b. A T.V. is on sale at "12% off."
       How much will be saved on a $400 set?

   c. Miss Brook borrowed $900. If she pays it back within a
       month she will only be charged 12% interest.
       How much will this interest payment be?

   d. Mrs. Dot's son sells cars. He gets to keep 9% of his
       sales. One week he had sales of $4500.
       How much was he allowed to keep?

   e. Mr. Knapp made a 16% down-payment on a sofa.
       How much was the down-payment if the sofa cost $450?

   f. "Big Spender" and his family eat out every Sunday. He
       always leaves a 15% tip. Last Sunday the bill was $60.
       How much was the tip?

2. Make up a problem which can be solved by using the one-percent
   method. Solve the problem. Give it to a classmate to solve.

PERCENT ALLOWANCE

1. Tom is a ninth-grader. He and his dad made a deal. Each month Tom gets 1% of his dad's $1200 paycheck.

   a. What is Tom's allowance for one month?
   b. Do you think Tom made a good deal?
   c. How does your allowance compare to Tom's?

   According to the agreement, Tom's allowance will increase 1% for each year he goes to school. Next year Tom's monthly allowance will be 2% of $1200. (We will pretend that Tom's dad will always bring home $1200.)

   d. How much money will he get as a tenth-grader?
   e. What percent will Tom be getting as a twelfth-grader?
   f. What will be his allowance as a twelfth-grader?
   g. What do you think of Tom's deal now?

   After high school, Tom plans to go to college.

   h. What will his allowance be during his third year in college?
   i. Will Tom need a part-time job?

2. Mary made a similar deal with her mother when she was in the ninth grade. Mary wants to be a doctor. It takes 8 years of college to become a doctor. Her mother's paycheck is $1500 each month. What will Mary's allowance be during her last year in medical school?

3. Suppose Mary's allowance was $225 her last year in medical school. How much would her mother's paycheck be?

Percent Allowance

Mathematics teaching objectives:

. Solve percent story problems using one-percent method.

Problem-solving skills pupils might use:

. Make necessary computations needed for the solution.

. Make explanations based upon data.

. Work backwards.

Materials needed:

. None

Comments and suggestions:

. Pupils use the one-percent method to solve the problems. Two pre-requisites are necessary:

  a. an understanding that 1% means one-hundredth
  b. skill in finding one-hundredth of a number (dividing by 100).

. Lots of discussion is possible regarding reasonable allowances, costs of higher education, etc.

Answers:

1. a. $12
    b. Answers will vary.
    c. Answers will vary.
    d. $24
    e. 4%
    f. $48
    g. Answers will vary.
    h. $84
    i. Probably

2. $180

3. $1875 (Working backwards from 12% to 1% is a good method to use and should be discussed.)

PERCENT APPLICATIONS

1. Spendless Department Store is offering a 15% discount on all items. A 15% discount means a reduction of 15¢ on every $1.00. Complete this 15% discount chart.

_15¢_ on every	$ 1.00	
___¢ on every	$ 2.00	
___¢ on	$ 3.00	
___¢ on	$10.00	
___¢ on	$21.00	

2. Complete this 18% discount chart.

_18¢_ on every	$ 1.00
___¢ on	$ 2.00
___¢ on	$ .50
___¢ on	$ 1.50
___¢ on	$ 6.50
___¢ on	$10.50

3. Some states have a sales tax of 6%. Complete this 6% tax table.

_6¢_ on every	$ 1.00
___¢ on every	$ 8.00
___¢ on	$25.00
___¢ on	$ .50
___¢ on	$22.50
___¢ on	$49.50

4. A bank pays an 8% interest rate. Complete this table.

$ 8 on every	$100
$ 4 on every	$____
$ 2 on every	$____
$16 on every	$____
$20 on every	$____
$46 on every	$____

5. Spendless is having a sale-- 12% off on all items. Complete this 12% table.

_12¢_ on every	$ 1.00
___¢ on every	$15.00
___¢ on every	$25.00
___¢ on every	$ .50
___¢ on every	$16.50

6. Estimate the answer to each problem below.

   a. A 12% discount on a $398 TV
   b. A 15% downpayment on a $1995 car
   c. A 13% interest payment on a $203 loan
   d. A 7% tax on a $498 stereo

Percent Applications

Mathematics teaching objectives:

. Solve percent problems involving discount, interest, and sales tax.

. Solve percent problems using the "on every hundred" method.

Problem-solving skills pupils <u>might</u> use:

. Make and use a systematic listing in a table.

. Identify patterns suggested by data.

. Make necessary computations needed for solution.

. Make reasonable estimates and answers.

Materials needed:

. None

Comments and suggestions:

. The "on every 100" method is closely related to the "ratio-proportion" method.

. Perhaps more problems like number 6 should be given so that pupils have more experience with the "on every hundred" method as well as making estimations. You may wish to use some "discount" adds from the newspaper.

. This lesson will take less than 1 period.

Answers:

1.   <u>15¢</u> on every $1.00
      <u>30¢</u> on every $2.00
      <u>45¢</u> on every $3.00
      <u>150¢</u> on every $10.00
      <u>315¢</u> on every $21.00

2.   <u>18¢</u> on every $1.00
      <u>36¢</u> on every $2.00
      <u>9¢</u> on every $ .50
      <u>27¢</u> on every $1.50
      <u>117¢</u> on every $6.50
      <u>189¢</u> on every $10.50

3.   <u>6¢</u> on every $ 1.00
      <u>48¢</u> on every $ 8.00
      <u>150¢</u> on every $25.00
      <u>3¢</u> on every $ .50
      <u>135¢</u> on every $22.50
      <u>297¢</u> on every $49.50

4. $ 8 on every $<u>100</u>
      $ 4 on every $<u>50</u>
      $ 2 on every $<u>25</u>
      $16 on every $<u>200</u>
      $20 on every $<u>250</u>
      $46 on every $<u>575</u>

5. $ .12 on every $ 1.00
      $1.80 on every $15.00
      $3.00 on every $25.00
      $ .06 on every $ .50
      $1.98 on every $16.50

6.   a. Approximately $ 48
      b. Approximately $300
      c. Approximately $ 26
      d. Approximately $ 35

USING PERCENT SENSE

Would you say that your percent sense has improved 100%?
Perhaps not--but it must be better than it was.

Many of the methods you have learned in this unit can be applied
to the problems below. In each case, decide which answer is the
most reasonable. Use good percent sense.

1. Sammy had 17 out of 21 correct. About what percent is this?

    a. 50%   b. 80%   c. 40%   d. 96%

2. Jodi made 8 baskets out of 17 tries. About what percent is
    this?

    a. 55%   b. 8%   c. 17%   d. 45%

3. Sherri had to pay 15% down. If the bill was $95, about how
    much was the downpayment?

    a. $50   b. $15   c. $1.50   d. $75

4. About what percent of the time would you
    expect to spin a "red?"

    a. 30%   b. 20%   c. 50%   d. 4%

    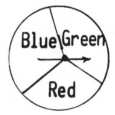

5. A 600-foot ditch is to be dug. It is 95% completed. About
    how much of the 600 feet has been completed?

    a. 95 feet   b. 450 feet   c. 550 feet   d. 300 feet

6. About what percent of this square has been shaded?

    a. 40%   b. 50%   c. 60%   d. 70%

    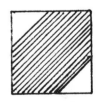

<u>Using Percent Sense</u>

Mathematics teaching objectives:
. Solve a variety of percent problems given in a multiple-choice format.
. Use estimation to get a reasonable answer.

Problem-solving skills pupils <u>might</u> use:
. Recognize limits and eliminate possibilities.
. Make necessary computations needed for solution.
. Make reasonable estimates as answers.

Materials needed:
. None

Comments and suggestions:
. Remind pupils that the possible choices for answers are only estimates.
. Have pupils discuss the various methods they used to arrive at their answers.
. This lesson will take less than 1 period.

Answers:

1. (b)
2. (d)
3. (b)
4. (a)
5. (c)
6. (d)
7. (b)
8. (c)
9. (b)
10. Answers will vary.

Using Percent Sense (cont.)

7. Bobbie had to pay a 5% sales tax. About how much tax should she pay on a $3999 car?

   a. $2000   b. $200   c. $20   d. $500

8. Jamie keeps 24% of what she collects on her paper route. If she keeps $30, about how much did she collect?

   a. $7   b. $30   c. $110   d. $60

9. Segment AB represents 32% of the distance between two towns. Which segment best represents the entire distance?

   A_____B

   a. _____   b. _____

   c. _____   d. _____

10. What percent do you think you've achieved on these 9 problems?

    a. 10%   b. 50%   c. 80%   c. 100%

Grade 7

VI. FACTORS, MULTIPLES, AND PRIMES

VI. FACTORS, MULTIPLES AND PRIMES

What is a perfect number?
Is "6" any more perfect than "10"?
Is "18" an abundant number?
What is so deficient about "23" ?

A study of certain number
theory topics can provide many
varied problem-solving opportunities. In this unit experiences are provided which
enable pupils to experiment,
explore, investigate, make
conjectures, test conjectures,
make generalizations and use
many other problem-solving skills.
In addition, pupils have many
opportunities for computation
practice in a problem-solving setting.

Using The Activities

It is best if the lessons are presented in the same sequence as given
in the unit.

A good way to introduce pupils to multiples and primes is to use what is
normally called the "Seive of Eratosthenes." (See WORLD BOOK ENCYCLOPEDIA,
1979 Edition.)

An excellent challenge activity suitable for 7th graders can be found in
the May, 1975 issue of the Arithmetic Teacher. The title of the article is
"A Crossnumber Game With Factors."

FACTORS

	Number	Set of Factors.
Example	20	2, 4, 10, 20, 1, 5
1.	18	9, 2, 18, 1, ___, ___
2.	12	___, ___, ___, ___, ___, ___
3.	30	___, ___, ___, ___, ___, ___, ___, ___
4.	35	___, ___, ___, ___
5.	___	3, 5, 1, ___
6.	___	4, 2, ___, ___
7.	___	19, ___
8.	___	37, ___
9.	___	5, 1, ___
10.	___	11, ___, ___
11.	___	7, ___, ___
12.	___	11, ___, ___, ___
13.	___	4, 14, ___, ___, ___, ___
14.	___	21, 2, ___, ___, ___, ___, ___, ___
15.	___	6, 10, ___, ___, ___, ___, ___, ___

16. Write the smallest number that has exactly

   a. 1 factor   c. 3 factors
   b. 2 factors   d. 5 factors

17. Felix, the class "trickster," said that there is no number that has exactly 7 factors. Can you show that Felix is wrong?

Research Questions

   . Why is 6 considered to be a Perfect Number?
   . What other perfect numbers can you find?
   . What is an abundant number?
   . Why is 23 deficient?

Factors

Mathematics teaching objectives:

    . Recognize and use properties of multiples and factors.
    . Use multiplication and division of whole numbers.

Problem-solving skills pupils _might_ use:

    . Guess and check.
    . Record solution possibilities or attempts.
    . Determine limits and/or eliminate possibilities.

Materials needed:

    . None

Comments and suggestions:

    . Pupils should work independently on this activity.

    . Some pupils may need to see one or two more examples.

    . Point out that the set of factors does not have to be listed in any particular order.

    . Two of the numbers in the list have only two factors. This might be a good time to identify these numbers as primes and have the class list others that would belong to this set.

    . Some pupils may observe that square numbers always have an odd number of factors. This property may be worthy of some discussion.

Answers:

	Number	Set of Factors
1.	18	9, 2, 18, 1, <u>3</u>, <u>6</u>
2.	12	<u>1</u>, <u>2</u>, <u>3</u>, <u>4</u>, <u>6</u>, <u>12</u>
3.	30	<u>1</u>, <u>2</u>, <u>3</u>, <u>5</u>, <u>6</u>, <u>10</u>, <u>15</u>, <u>30</u>
4.	35	<u>1</u>, <u>5</u>, <u>7</u>, <u>35</u>
5.	<u>15</u>	3, 5, 1, <u>15</u>
6.	<u>8</u>	4, 2, <u>1</u>, <u>8</u>
7.	<u>19</u>	19, <u>1</u>
8.	<u>37</u>	37, <u>1</u>
9.	<u>25</u>	5, 1, <u>25</u>
10.	<u>121</u>	11, <u>1</u>, <u>121</u>
11.	<u>49</u>	7, <u>1</u>, <u>49</u>
12.	<u>22</u>	11, <u>1</u>, <u>22</u>, 2
13.	<u>28</u>	4, 14, <u>1</u>, <u>2</u>, <u>7</u>, <u>28</u>
14.	<u>42</u>	21, 2, <u>1</u>, <u>3</u>, <u>6</u>, <u>7</u>, <u>14</u>, <u>42</u>
15.	<u>30</u>	6, 10, <u>1</u>, <u>2</u>, <u>3</u>, <u>5</u>, <u>15</u>, <u>30</u>

(Note beside items 12–14: This is one of an unlimited number of possibilities. Pupils could be encouraged to find others.)

16. a. 1   b. 2   c. 4   d. 16

17. 2^6 or 64 has exactly seven factors. So do 3^6, 5^6, 7^6, 11^6, 13^6, ...

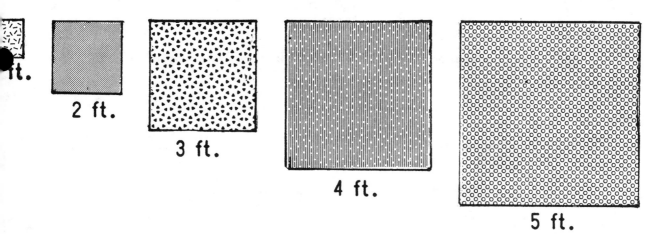

1 ft.
2 ft.
3 ft.
4 ft.
5 ft.

Mrs. Gonzales has found a place to buy sections of square carpeting. The pieces come in five different patterns and five different sizes.

1. It's easy to see how Mrs. Gonzales could cover the sewing room floor by using 1-foot tiles.

   Is it possible to cover the floor with pieces 2 feet on a side (without cutting)? Explain.

8 ft.

4 ft.

Sewing Room

2. Which of the other pieces could be used to exactly cover the floor?

3. Which of the carpet sizes could be used to cover the floor of the following rooms? Record your answers in the chart.

Room	Sizes that exactly cover	Largest Size
Sewing 4 ft. by 8 ft.		
Closet 6 ft. by 9 ft.		
Pantry 10 ft. by 4 ft.		
Bedroom 10 ft. by 15 ft.		
Kitchen 16 ft. by 12 ft.		
Workroom 7 ft. by 12 ft.		
Living 16 ft. by 20 ft.		
Dining 11 ft. by 17 ft.		

Common Factors

Mathematics teaching objectives:

  . Determine the common factors and greatest common factor of a pair
    of numbers.

Problem-solving skills pupils *might* use:

  . Use a drawing or model.
  . Work backwards.
  . Create problems similar to ones already solved.

Materials needed:

  . Small squares cut from tagboard (optional)
  . Scale models of rooms. (optional)

Comments and suggestions:

  . Pupils may need some assistance with problem 7. Therefore, it is
    probably best that the whole activity be done during class.

  . Pupils who have difficulty with visualization could actually use small
    tagboard squares and scale models of the rooms.

  . No attempt has been made to relate these "common factor" experiences
    to fractions. You may wish to do this.

Answers:

  1. Yes. Two pieces will fit along the width and four along the length
      (a total of 8 pieces).

  2. 4 by 4

  3.

Room		Sizes that exactly cover	Largest size
Sewing Room	4 ft. by 8 ft.	1, 2, 4	4
Closet	6 ft. by 9 ft.	1, 3	3
Pantry	10 ft. by 4 ft.	1, 2	2
Bedroom	10 ft. by 15 ft.	1, 5	5
Kitchen	16 ft. by 12 ft.	1, 2, 4	4
Workroom	7 ft. by 12 ft.	1	1
Living Room	16 ft. by 20 ft.	1, 2, 4	4
Dining Room	11 ft. by 17 ft.	1	1

Common Factors (cont.)

4. Examine the completed chart.

    a. What size works for all the floors?
    b. How can you tell whether a size will work or not?

5. What size would cover these floors exactly? (You may use other square pieces like 6 by 6, 7 by 7, 8 by 8, etc.)

    a. 8 ft by 16 ft   d. 15 ft by 25 ft
    b. 10 ft by 20 ft   e. 14 ft by 21 ft
    c. 18 ft by 24 ft   f. 9 ft by 14 ft

The sizes that work
are divisors of both
numbers.
These divisors are
called
COMMON FACTORS.

The largest size that
works is called the
GREATEST COMMON FACTOR.

6. Find the common factors of each pair of numbers. Then find the greatest common factor of each pair.

    a. 15 and 21   d. 16 and 24
    b. 40 and 50   e. 24 and 36
    c. 30 and 45   f. 18 and 27

7. In problem 6 you were given pairs of numbers and asked to find the common factors. Let's see if it's possible to go the other way. Complete as much of the chart as possible.

	Pairs of Numbers	All Common Factors
a.		1 and 3
b.		1 and 5
c.		1, 2, and 3
d.		1, 2, and 4
e.		1, 3, and 9
f.		1, 3, and 5

Common Factors

4. a. 1
    b. If the size is a factor of both dimensions of the room.

5. a. 1, 2, 4, 8   d. 1, 5
    b. 1, 2, 5, 10   e. 1, 7
    c. 1, 2, 3, 6   f. 1

6. a. 1, 3   d. 1, 2, 4, 8
        GCF is 3   GCF is 8

    b. 1, 2, 5, 10   e. 1, 2, 3, 4, 6, 12
        GCF is 10   GCF is 12

    c. 1, 3, 5, 15   f. 1, 3, 9
        GCF is 15   GCF is 9

7. a. Some possibilities are   d. Some possibilities are
            3 and 6   4 and 8
            3 and 9   4 and 12
            6 and 9   8 and 12

    b. Some possibilities are   e. Some possibilities are
            5 and 10   9 and 18
            5 and 15   9 and 27
            10 and 15   18 and 27

    c. Not possible   f. Not possible

8. Answers will vary. For part (b), if the given "factors"
    (except for 1) are prime, then the problem is impossible.
    For example: 1, 5, 7.

Common Factors (cont.)

8. Some of the problems in number 7 were possible to solve; some were impossible to solve.

    a. Make up two more problems, like those in number 7, that are possible to solve.

    b. Make up two that are impossible to solve.

COMMON MULTIPLES

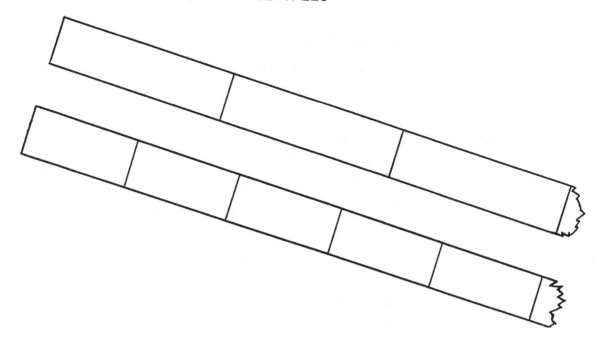

1. Jill and Jim each started with 30 cm of adding machine tape.
   Jill cut hers into 5 cm pieces. Jim cut his into 3 cm pieces.

   a. How many pieces did Jill have?

   b. How many pieces did Jim have?

2. Suppose they started with 18 cm strips.

   a. Is it still possible to cut the strips into 3 cm and
       5 cm pieces without any waste? Explain.

   b. What lengths are possible to begin with?
       Give at least 4 other possibilities.

Common Multiples

Mathematics teaching objectives:

. Determine common multiples and the least common multiple of a
  set of numbers.

Problem-solving skills pupils *might* use:

. Use a drawing or model.

. Make generalizations based upon data.

Materials needed:

. Adding machine tape (optional)

Comments and suggestions:

. No attempt has been made to relate these "common multiple"
  experiences to fractions. You may wish to do this.

. You may wish to provide some pupils with strips of adding
  machine tape so that they can actually perform the experiment.

Answers:

1. a. 6
    b. 10

2. a. No; 3 is a factor of 18 but 5 is not.

    b. Any multiple of 15

(Answers continued on next page...)

Common Multiples (cont.)

3. Jill and Jim want to cut other strips of paper tape into pieces.
    Complete the table below. Note that the example just completed
    is shown in the first row.

Size of Pieces	List of Four Possible Lengths	Smallest Possible Length
3 cm and 5 cm	15, 30, 45, 60	15
2 cm and 7 cm		
4 cm and 6 cm		
4 cm and 8 cm		
3 cm and 12 cm		
6 cm and 9 cm		
6 cm and 10 cm		

4. Refer to the chart in problem 3.
    The numbers in the middle column are called Common Multiples.
    The number in the last column is the Least Common Multiple.

    a. List four common multiples of 15 and 20.
    b. What is the least common multiple of 15 and 20?

5. List four common multiples of these sets of numbers and
    the least common multiple of each set.

    a. 3, 4, 12   d. 2, 3, 5
    b. 2, 3, 4   e. 2, 4, 6
    c. 5, 10, 15   f. 3, 4, 8

6. Find the least common multiple of 2, 3, 4, 5, 6.

7. What five numbers have a least common multiple of 20?

Answers: (cont.)

3.

Size of Pieces	List of Four Possible Lengths	Smallest Possible Length
3 cm and 5 cm	15, 30, 45, 60	15
2 cm and 7 cm	any multiple of 14	14
4 cm and 6 cm	" " " 12	12
4 cm and 8 cm	" " " 8	8
3 cm and 12 cm	" " " 12	12
6 cm and 9 cm	" " " 18	18
6 cm and 10 cm	" " " 30	30

4. a. Any multiple of 60

    b. 60

5. a. Any multiple of 12   d. Any multiple of 30
       LCM is 12   LCM is 30

    b. Any multiple of 12   e. Any multiple of 12
       LCM is 12   LCM is 12

    c. Any multiple of 30   f. Any multiple of 24
       LCM is 30   LCM is 24

6. 60

7. 1, 2, 4, 5, 10   or   2, 4, 5, 10, 20

PRIME NUMBERS

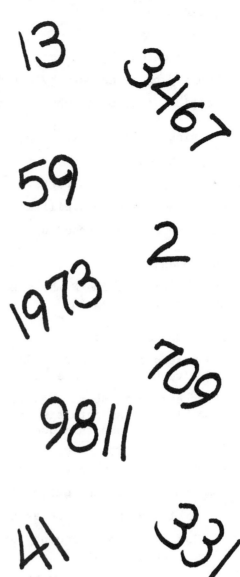

How can a person (or better yet, a computer) determine whether a number is prime? It's easy to show that 37 and 53 are prime. But how about 289 or 739?

Let's see if we can discover a method.

1. Look at the numbers below. Which of them are you absolutely certain are <u>not</u> prime? How do you know?

120	882	531
289	367	827
355	49	99

2. Is 289 prime? Let's find out. Our main goal is to see if we can find any divisors of 289.

   a. Why is it unnecessary to divide by 2? How about 4 or 6 or other multiples of 2?

   b. Why is it unnecessary to divide by 5? How about 10 or 15 or other multiples of 5?

   c. Is it necessary to divide by 3? If so, show the division.

   d. List numbers you think should be tried as divisors. Then show the divisions.

   e. Is 289 a prime number? Why?

Prime Numbers

Mathematics teaching objectives:

. Identify prime numbers by divisibility tests and by using a chart.

. Practice division of whole numbers.

Problem-solving skills pupils _might_ use:

. Guess and check.

. Break a problem into parts.

. Record solution possibilities or attempts.

. Make explanations based upon results.

Materials needed:

. Sheet titled "Prime Numbers Less Than 1000" (to be handed out near the end of the lesson)

Comments and suggestions:

. The lesson should be done during class so that you can react to questions and give hints when needed.

. The question in 3b may be difficult for pupils to answer, although they probably have an intuitive feeling as to why 23 will not work. Perhaps you could discuss this question with the entire class at the appropriate time.

. Do _not_ give a pupil the list of primes until problem 4 has been completed.

. Some possible investigations for 5b:

. Is there any pattern as to the frequency of primes?

. How about the frequency of prime twins, like 41 and 43, 71 and 73, etc. ?

. Do primes end in 1 more frequently than in 7?

Answers:

1. Answers will vary. Hopefully, pupils will know from previous experience that

120	is divisible	by	10	(ends in "0")
355	"	"	5	(ends in "5")
882	"	"	2	(is even)
49	"	"	7	(is a multiple of 7)
99	"	"	11	(is a "double-digit" number)

(Answers continued on the next page...)

Prime Numbers (cont.)

3. Is 367 a prime number?

   a. List the first 6 numbers you think you should try as divisors. Then show the divisions.

   b. Why is it unnecessary to try 23 as a divisor?

   c. Is 367 a prime number?

4. Determine whether each of these numbers is prime. If it is not prime, write down the smallest divisor (other than 1).

   a. 171   b. 223   c. 401

5. Get a sheet from your teacher called "Prime Numbers Less Than 1000."

   a. Check your answers to problem 4.

   b. Can you find any patterns in the list? If so, describe them.

Answers: (cont.)

2.   a.   If you were to count by 2's, or by 4's, or by 6's, etc.,
          the result would always be an even number.

     b.   If you were to count by 5's, or by 10's, or by 15's, etc.,
          the result would either end in a "0" or a "5".

     c.   Yes - unless the pupils know the test for divisibility by 3.
          $$289 \div 3 = 96 \text{ R1}$$

     d.   7 should be tried; $289 \div 7 = 41 \text{ R2}$

          11 should be tried; $289 \div 11 = 26 \text{ R3}$

          13 should be tried; $289 \div 13 = 22 \text{ R3}$

          17 should be tried; $289 \div 17 = 17 \text{ R0}$

     e.   289 is <u>not</u> prime. $289 = 17 \times 17$

3.   a.   Answers will vary. A good set of divisors to try is:
          3, 7, 11, 13, 17 and 19.

          $$367 \div 3 = 122 \text{ R1}$$
          $$367 \div 7 = 52 \text{ R3}$$
          $$367 \div 11 = 33 \text{ R4}$$
          $$367 \div 13 = 28 \text{ R3}$$
          $$367 \div 17 = 21 \text{ R10}$$
          $$367 \div 19 = 19 \text{ R6}$$

     b.   Notice that as the divisor (one factor) increases, the
          quotient (the other "factor") decreases. If 23 is a factor
          then the other factor would have to be less than 19. This
          is impossible since all possible factors less than 19 have
          already been tried.

     c.   Yes.

4.   a.   Not prime   (3)   b.   Prime   c.   Prime

5.   Answers will vary.

PRIME NUMBERS LESS THAN 1000

2	3	5	7	11
13	17	19	23	29
31	37	41	43	47
53	59	61	67	71
73	79	83	89	97
101	103	107	109	113
127	131	137	139	149
151	157	163	167	173
179	181	191	193	197
199	211	223	227	229
233	239	241	251	257
263	269	271	277	281
283	293	307	311	313
317	331	337	347	349
353	359	367	373	379
383	389	397	401	409
419	421	431	433	439
443	449	457	461	463
467	479	487	491	499
503	509	521	523	541
547	557	563	569	571
577	587	593	599	601
607	613	617	619	631
641	643	647	653	659
661	673	677	683	691
701	709	719	727	733
739	743	751	757	761
769	773	787	797	809
811	821	823	827	829
839	853	857	859	863
877	881	883	887	907
911	919	929	937	941
947	953	967	971	977
983	991	997		

SOME PRIME INVESTIGATIONS

1. You'll need a "Prime Numbers Less than 1000" sheet.

   > Every even number (greater than 4) can be written as the sum of two odd primes.

   $10 = 7 + 3$

   $12 = 7 + 5$

   $14 = 7 + 7$

   $16 = 11 + 5$

   a. Do you think this statement is always true? Try some numbers to see.

   b. Do you think the statement is true if the two primes must be different? Try some numbers to see.

2. Goldbach was a famous Russian mathematican. His most famous discovery is the one you experimented with above. Everyone would like to become famous sometime during their life. Arthur Hamann, a 7th grade student in Illinois, succeeded.

   ## Hamann's conjecture

   MAXINE R. FRAME

   The Hamann of the title was a student in the seventh grade class of the author, Maxine Frame, who teaches junior high school mathematics at the Horace Mann School in Oak Park, Illinois.

   From "Arithmetic Teacher," January 1976.

   > "Every even number is the difference between two primes."
   >
   > Hamann's Conjecture

   a. Do you think that Hamann's statement is always true? Try some numbers to see. For example, $16 = 23 - 7$, $24 = 29 - 5$.

   b. How many different solutions can you find for 16? for 24?

Some Prime Investigations

Mathematics teaching objectives:

. Test a variety of conjectures from number theory.

. Practice whole number computation.

. Evaluate a formula.

Problem-solving skills pupils _might_ use:

. Search printed matter for needed information.

. Record solution possibilities or attempts.

. Look for patterns.

. Look for counter-examples of mathematical conjectures.

Materials needed:

. "Prime Numbers Less Than 1000" sheet

Comments and suggestions:

. This lesson should be done during class so that you can interact while pupils are involved in the investigations.

. You should discuss the importance of conjectures (hypotheses) and the role they play in mathematical investigation.

. The term "counter-example" should be emphasized when discussing pupils' attempts to find cases that do not fit a hypothesis.

Answers:

1. a. Answers will vary. No counter-examples have ever been found. Nevertheless, the statement has never been proven mathematically.

   b. Answers will vary. Once again, no counter-examples have ever been found.

2. a. No counter-examples have been found.

   b. There appear to be an unlimited number of ways of representing any even number using Hamann's conjecture.

$16 = 19 - 3$	$24 = 29 - 5$
$16 = 23 - 7$	$24 = 37 - 13$
$16 = 29 - 13$	$24 = 41 - 17$
$16 = 47 - 31$	$24 = 43 - 19$
$16 = 53 - 37$	$24 = 47 - 23$

(Answers continued on next page.

Some Prime Investigations (cont.)

3. 29 and 31 are prime twins. So are 41 and 43.

> The "middle number" of prime twins can always
> be divided by 6.

Do you think that this statement is always true?
Try some numbers to see.

4. Beth discovered a formula which she says will always give
a prime number.

$$P = N \times N + N + 17$$

Will Beth's formula always give a prime? Complete the table
and extend it to N = 17.

N	0	1	2	3	4	5	
P	17	19	23	29			. . .

5. Jeff made a statement about primes.

> There are nine 2-digit primes that are also
> prime when the digits are reversed.

Do you think Jeff's statement is correct? See how many
you can find. Two of them are 17 and 71.

EXTENSION

Perhaps you can make a prime discovery of your own. Try it.
Who knows, maybe you'll become famous.

Some Prime Investigations

3. No counter-examples have been found, except for the prime twins, 3 and 5.

4.

N	0	1	2	3	4	5	6	7	8	9	10	11	12	13	14	15	16	17
P	17	19	23	29	37	47	59	73	89	107	127	149	173	199	227	257	289	323

No - it breaks down at N = 16 and N = 17.

5. Jeff is correct.

11,   13,   17,   31,   37,   71,   73,   79,   97

Extension:

Answers will vary.

Grade 7

"Big Willie" has huge feet. He was
curious as to just how big they actually
were. He figured he might be able to find
the area of his foot by using string.

. First, Willie cut a piece of string
   the same length as the perimeter
   (distance around) of his foot.

. Next, he used that length of string to make a square.

. Finally, he found the area enclosed by the square.

Do you think that Willie's method is reasonably accurate?*

Many people are certain that "Big Willie's" method works, until they try it.
They feel that by keeping the perimeter constant, the area will also remain fixed,
and visa versa. The same faulty thinking often occurs in problems which relate
surface area and volume. Several of the problem situations in this unit are
designed to make pupils more aware of these relationships.

Generally speaking, this unit is not intended to develop the concepts of
perimeter, area, and volume. This needs to be done earlier. Instead, the unit
should be used to reinforce these measurement concepts and to show some applications.

Several lessons provide excellent opportunities to illustrate strategies
which make use of the skill, "Make a systematic listing," one of the more
difficult problem-solving skills for pupils to master. On occasion, some
teacher direction may need to be supplied when the solution depends upon a
systematic listing.

Using The Activities

The activities in the unit can be done in any order. You may wish to
provide additional activities, especially those that are situations drawn from
the pupils "real world" of measurement.

Materials needed: . Two-centimetre or inch cubes
                     . Tape measures
                     . Mail-order catalogs
                     . Supply of notebook-size tagboard
                     . Scissors
                     . Centimetre rulers
                     . Calculators

*From MATHLAB-JUNIOR HIGH, Action Math Associates, Inc.

STACKING BLOCKS

This figure has a volume of 2 cubic centimetres.
It has a surface area of 10 square centimetres.

Determine the volume and surface area of each
figure below.

1.

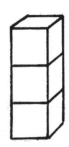

2.

3.

4.

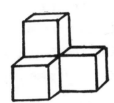

5.

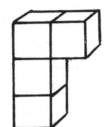

6.

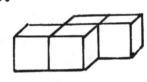

7.

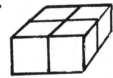

8.

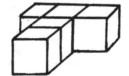

9.

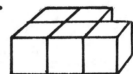

10.

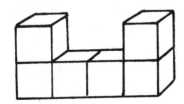

11.

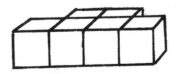

EXTENSION

Draw a shape using 6 blocks that has a surface area less
than 24.

Stacking Blocks

Mathematics teaching objectives:

    . Find surface area and volume of "block" figures by counting.

    . Make a drawing of a "block" figure with a given surface area.

Problem-solving skills pupils _might_ use:

    . Use a physical model.

    . Make predictions and conjectures based upon observed patterns.

    . Look at problem situations from varying points of view.

    . Search for and/or be aware of other solutions.

Materials needed:

    . A set of cubes should be available for class demonstration and for those pupils who have difficulty visualizing drawings of three-dimensional figures.

Comments and suggestions:

    . Pupils may need some review of the concepts of surface area and volume.

    . After several examples are solved by actually building the objects, pupils should be encouraged to find answers for the remaining examples without using the cubes. They then could check their answers by building the objects and then counting the squares and cubes.

    . This activity will take about 15 minutes to complete. It and the next activity, "Cubic Inches," can be completed in one class period.

Answers:

	1	2	3	4	5	6	7	8	9	10	11
Volume	3	3	4	4	4	4	4	5	5	6	6
Surface area	14	14	18	18	18	18	16	22	20	26	24

A careful study of examples 3, 4, 5, and 6 suggests ways example 8 could be varied to draw other "block" objects with a surface area of 22 and a volume of 5.

Extensions:

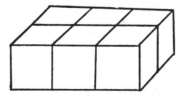

Shape has a volume of 6 and a surface area of 22.

CUBIC INCHES

"Billie the Bandit" was serving his time in jail. He had held up the northbound stage. It's easy for him to recall the incident.

Billie had told the driver to throw down the box of gold. It wasn't very large--only 6 inches by 4 inches by 2 inches. Billie found out later that 1 cubic inch of gold weighs approximately 2.7 pounds.

Why do you suppose Billie was caught? Work through the class exercises to find out.

Class Exercises

1. How many cubic inches could be placed in one layer of the box?

2. How many cubic inches would fill the box?

3. If the box was filled with gold, how much would the gold weigh?

4. Why did Billie get caught?

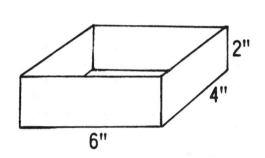

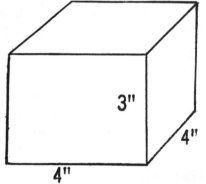

Volume of 48

Length	Width	Height
6	4	2
4	4	3

Exercises

1. The volume of the box at the right is also 48 cubic inches. In fact, there are nine different boxes that have a volume of 48 cubic inches. Find all nine of them and place the dimensions in the table. The first two are listed for you.
(Note: A 4 by 4 by 3 box, a 3 by 4 by 4 box, and a 4 by 3 by 4 box are all the same.)

Cubic Inches

Mathematics teaching objectives:

. Given the dimensions of a box, calculate its volume.

. Fine combinations of three factors which have a given two-digit product.

Problem-solving skills pupils might use:

. Determine measurements for obtaining a solution.

. Record solution possibilities.

. Search for and be aware of other solutions.

Materials needed:

. None

Comments and suggestions:

. Use the introductory setting as a way of getting into the lesson. Let pupils make conjectures as to why Billie was caught. Work the Class Exercises via class discussion. At this time, the rationale for the volume formula could be reviewed or developed.

. Let pupils work the Exercises independently as you offer assistance where needed.

. A discussion of the lesson should include a listing of the different possible whole-number dimensions for a box with a volume of 48.

. This activity will take about 15 minutes to complete. It and the previous activity, Stacking Blocks, can be completed in one class period.

Answers:

Class Exercises

1. 24   2. 48   3. 129.6 lbs.

4. The box of gold was much heavier than anticipated.
   Billie may have had difficulty even lifting the box.

Exercises

1. There are 9 different boxes with whole-number dimensions
   and with a volume of 48 cubic inches.

48, 1, 1	12, 4, 1	8, 2, 3
24, 2, 1	12, 2, 2	6, 4, 2
16, 3, 1	8, 6, 1	4, 4, 3

Possible
Extension: How many different-sized boxes will hold 72-inch cubes?

96-inch cubes?

(Answers continued on next page...)

Cubic Inches (cont.)

2. Which of the boxes listed do you think would be most appropriate for boxing up cubes? Explain.

3. Suppose a box was completely filled with 47 one-inch cubes. List all possible dimensions for a box such as this.

EXTENSION

Refer back to exercise 3.

What other numbers less than 100 could be substituted for 47 to give the same number of possibilities.

Answers:

Exercises (cont.)

2. Permit any answer which has a plausible explanation. A 48 by 1 by 1 box would be easier to hold firmly with one hand. An 8 by 6 by 1 box would easily fit on a shelf with books. After several succeeding lessons, pupils should recognize that a 4 by 4 by 3 box could be made with the least amount of material. This reason shouldn't be given unless it's suggested by a pupil.

3. Only one possibility: 47 by 1 by 1 box.

Extension: All the prime numbers less than 100. A 1 by 1 by 1 box will also work.

BOXING UP CUBES

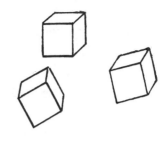

Kenny and Mike decided to start a part-time business--manufacturing one-inch cubes for classroom use. Now they need to decide how to box them up.

Each box is to contain 100 cubes. The problem is-- which size box will use the smallest amount of cardboard?

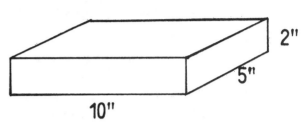

Kenny drew a picture like the one above. Will it hold 100 cubes?

1. How much cardboard is needed? (In other words, what is the total area?) Don't forget to include the top.

2. Determine the dimensions of all boxes that have volumes of 100. Also, find the total surface area of each. (Do not use decimals.) Place your results in the table below.

Volume	Dimensions	Total Surface Area
100	5 by 10 by 2	
100		
100		
100		
100		
100		
100		
100		
100		
100		

3. Which size box will use the smallest amount of cardboard?

EXTENSION   Suppose you wanted to design a box that would hold 100 cubic centimetres of cereal. Determine the dimensions that use the least amount of cardboard. Use decimals. A calculator would be helpful.

Boxing Up Cubes

Mathematics teaching objectives:

- Given the dimensions of a box, find the surface area.
- If the volume of a rectangular solid is fixed, determine the dimensions of the solid which will give the smallest surface area.

Problem-solving skills pupils _might_ use:

- Visualize an object from its drawing.
- Record solution possibilities in a table.
- Make conjectures based upon observed patterns.

Materials needed:

- Calculator (optional)

Comments and suggestions:

- You may need to review the concepts of volume and surface area.
- If pupils have completed the first two lessons they should be able to work this lesson independently by following the directions given in the introductory setting and in each problem.

Answers:

1. 160 square inches

2. All possible whole-number dimensions for boxes with a volume of 100 are given.

Volume	Dimensions	Total Area	Volume	Dimensions	Total Area
100	5 by 10 by 2	160 sq.in.	100	1 by 5 by 20	250 sq.in.
100	1 by 1 by 100	402 sq.in.	100	1 by 10 by 10	240 sq.in.
100	1 by 2 by 50	304 sq.in.	100	2 by 2 by 25	208 sq.in.
100	1 by 4 by 25	258 sq.in.	100	4 by 5 by 5	130 sq.in.

3. 4 by 5 by 5. Note that this box comes closest to being a cube than the other boxes. A good follow-up problem is to have pupils decide what dimensions would be best if the given volume is 64.

    Answer: A 4 by 4 by 4 cube - surface area of 96.

Extension: A reasonable size: 4.65" by 4.65" by 4.65"; amount of cardboard - 129.735 sq.in. This box has a volume slightly over 100 cu.in.

FLOOR TILES

Suppose you were to put new tile on the floor in your classroom.

How would you go about determining the cost?

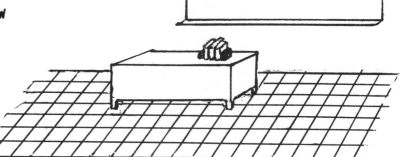

1. Write down the information you would need to know. Then discuss this with your teacher.

2. Now take the necessary measurements and determine the total cost.

EXTENSION

Suppose you were to put wall-to-wall carpeting on the floor. How much more would this cost?

PSM 81

Floor Tiles

Mathematics teaching objectives:

    . Collect the necessary data, take the necessary measurements, and determine the cost of tiling the classroom floor.

Problem-solving skills pupils might use:

    . Determine and collect data needed to solve the problem.

    . Listen to persons who have relevant knowledge and experiences to share.

    . Make a drawing.

    . Make decisions based upon data.

Materials needed:

    . 5 or 6 tape measures

    . Mail-order catalogs

Comments and suggestions:

    . This lesson has the most value if pupils research the problem themselves.

    . Possibly the class could be divided into brainstorming groups for sharing ideas before the problem is discussed as a total class.

    . Different groups of pupils could have different tasks - some could measure, some could record measurements on floor plan sketches, others could find prices in catalogs, and still others could contact local stores selling tile.

    . Different groups could work the problem and then compare their results.

Answers:

Answers will vary, depending on the size of the room and cost of material. An example is given below.

              Room size - 35 feet by 35 feet

              Area of floor - 1225 square feet

              Cost per tile - $ .79

              Total cost - 1050 x $ .79 = $829.50

A PEN FOR BARNEY

Barney needs a new dog pen. But Stephen Brown only has 48 feet of fencing to use. Of course, Stephen and Barney want the pen to have as much area as possible. Let's see if we can help them out.

1. Use squared paper. Draw a rectangle with a width of 1 and a perimeter of 48.

2. Draw other rectangular pens that use 48 feet of fencing. Record the dimensions and areas in the table.

3. Which rectangular pen is most suitable for Barney?

4. Suppose Stephen Brown has 40 feet of fencing, rather than 48 feet. Now what dimensions would give the most suitable pen for Barney?

Width	Length	Area
1	23	

EXTENSION

Suppose Stephen Brown used a side of his house as one side of the pen. Use 48 feet of fencing. Now what dimensions would give the greatest area? Use squared paper. Make a chart of your results.

Mathematics teaching objectives:

. Determine the dimensions of several rectangles with a given perimeter.

. If the perimeter of a rectangle is fixed, determine the dimensions which will give the greatest area.

Problem-solving skills pupils <u>might</u> use:

. Make and use a drawing.

. Make a systematic listing in a table.

. Identify trends (patterns) suggested by data in a table.

Materials needed:

. Squared paper (cm)

Comments and suggestions:

. The systematic listing of possible solutions seems to be a skill pupils are slow to acquire. This activity provides an excellent opportunity for showing the power of this problem-solving skill.

. Provide direction until pupils begin to see a pattern evolving and then let them complete the problem on their own.

. Provide class time for observing the patterns in the table shown below.

Answers:

2.

Width	Length	Area
1	23	23
2	22	44
3	21	63
4	20	80
5	19	95
6	18	108
7	17	119
8	16	128
9	15	135
10	14	140
11	13	143
12	12	144
13	11	143
14	10	140
15	9	135
⋮	⋮	⋮

3. 12' by 12'

Pupils should be encouraged to use the table as evidence for showing this answer is correct.

4. 10' by 10'

Perhaps even more examples will be needed before pupils discover that the largest area is a square.

Extension:

12' by 24' gives an area of 288 square feet.

GREATEST VOLUME

The Pauper Paper Company has been given the job of manufacturing boxes. The boxes are to be made from a square 18 cm by 18 cm. Also, they are to be open at the top.

This diagram illustrates how to construct a box from a piece of paper.

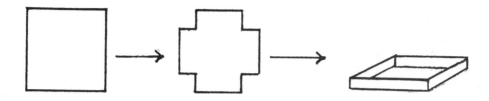

. Get squared paper and scissors.
. Cut out a square 18 cm by 18 cm.
. Cut a 1 cm by 1 cm square from each corner.
. Fold up the sides to make an open box.

1. What are the dimensions and the volume of the box you constructed? Place this information in the first row of the table.

2. Use another section of 18 cm by 18 cm paper. Make a box which has a greater volume than your first one. Place the dimensions and volume of your second box in the table.

Length	Width	Height	Volume

3. The Pauper Paper Company would like to find dimensions which give the <u>greatest</u> volume. See if you can help them out. Be sure to place your results in the table. If you use decimals, you may want to borrow a calculator.

Greatest Volume

Mathematics teaching objectives:

- Determine the dimensions of the lidless box (with the greatest volume) that can be made from a square of given size.

Problem-solving skills pupils __might__ use:

- Make and use a physical model.
- Guess and check.
- Identify trends (patterns) suggested by data in a table.

Materials needed:

- Classroom set of tagboard, 18 cm by 18 cm, marked in 1 cm squares
- Scissors
- Centimetre rulers
- Calculators (optional)

Comments and suggestions:

- Provide some direct instruction for the making of lidless box from a square, 18 cm by 18 cm.
- Elicit from the pupils the procedures for determining the volume of the box.
- Provide direction, if needed, in getting pupils started in making a systematic listing of "solutions" but let them complete the listing on their own.

Answers:

1.- 3.

Length	Width	Height	Volume
16	16	1	256
14	14	2	392
12	12	3	432
10	10	4	400
8	8	5	320
6	6	6	216

Note:

If pupils question whether or not the box with the indicated dimensions actually gives the greatest volume, suggest that they determine the volume of two boxes, one with a height of 2.9 cm and the other with a height of 3.1 cm and compare with the volume for the 3 cm height.

Length	Width	Height	Volume
12.2	12.2	2.9	431.636
11.8	11.8	3.1	431.644

Possible Extension:

What is the greatest volume of a lidless box that can be made from a 30 cm by 30 cm square?

Answer: The dimensions of the box with the greatest volume are 20 cm by 20 cm by 5 cm. ($V = 2000$ cm^3)

If more examples are given, some pupils may discover that the greatest volume will result when the height is $\frac{1}{6}$ of the side of the original square.

DESIGNING OFFICE SPACES

Miss Andrews is an architect. She's been asked to design an office building with several offices on each floor. Each individual office on the first floor is to be 72 square feet.

Class Exercises

1. Use squared paper. Draw a rectangle that is 2 by 36.

    a. What is the area of this rectangle?

    b. Do you think Miss Andrews would design rooms with these dimensions? Why.

2. Draw a rectangle that is 3 by 24.

    a. What is the area of this rectangle?

    b. Which would have more wall space,

    a 2 by 36 room   or   a 3 by 24 room?

    c. Miss Andrews wants to design rooms that will cost as little as possible to build. Does the perimeter of a room have an effect on the cost? Explain.

3. Draw other rectangles that have areas of 72. Find the perimeter of each rectangle. Fill in the chart.

Areas of 72

Width	Length	Perimeter
2	36	76
3	24	

4. Which dimensions are the best to use? Explain.

Mathematics teaching objectives:

. Solve problems involving perimeter and area.

. Give examples to show that rectangles with the same area can have different perimeters.

Problem-solving skills pupils might use:

. Make and use a drawing.

. Identify trends suggested by data in a table.

. Make explanations based upon data.

Materials needed:

. Squared paper

Comments and suggestions:

. Use the Class Exercises as a group activity, eliciting reponses as often as possible. Such feedback should give an indication of the effectiveness of the previous activities.

. Let pupils proceed individually on the Exercises.

Answers:

Class Exercises

1. a. 72 sq. ft.
   b. Probably not. The room would be too narrow even for a hallway.

2. a. 72 sq. ft.
   b. 2 by 36 room (assuming the heights of both rooms are the same).
   c. Yes. An increase in the perimeter of a room results in an increase in its wall space. An increase in wall space in this office building results in an increase in cost.

3.

Areas of 72

Width	Length	Perimeter
2	36	76
3	24	54
4	18	44
6	12	36
8	9	34
9	8	34

4. An 8 by 9 room probably would be preferable for the usual office with a desk, filing cabinets, typewriter, and chairs. As to building costs, the 8 by 9 room probably would be less expensive than rooms with the other dimensions.

(Answers continued on next page...)

Designing Office Spaces (cont.)

Exercises

1. Each of the offices on the second floor is to be 64 square feet. What dimensions would be best?

2. Each of the offices on the third floor is to be 120 square feet. Which dimensions would be best?

3. Miss Andrews also was asked to design a rectangular lunchroom with an area of 800 square feet. What dimensions would be best?

Answers: (cont.)

Exercises

1. 8 ft. by 8 ft. if the least wall space is the criteria.

2. 10 ft. by 12 ft. if the least wall space is the criteria.
   (If decimals are used, then the answer would be approximately
   10.95 ft. by 10.95 ft.)

3. Probably 25 ft. by 32 ft. if the least wall space is the
   criteria. (If decimals are used, the answer would be
   approximately 28.28 ft. by 28.28 ft.) If lots of window
   space is a criteria, then other dimensions may be more
   desirable.

SKYSCRAPER

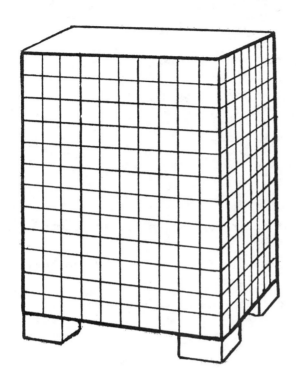

A calculator will be helpful in solving these problems.

1. This building is entirely covered by windows on all four sides.

   a. How many windows are there?

   b. Each window costs $88.44. How much did all the windows cost?

2. Once a year all the windows are washed. The costs are

   $2 for each 1st floor window.
   $2.50 for each 2nd floor window.
   $3 for each 3rd floor window.
     etc.
   How much does it cost to have all the windows washed?

Floor Plan (same for every floor)

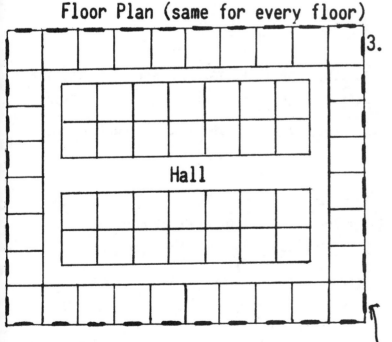

Hall

↳windows

3. All inside offices (those without windows) rent for $125 per month.
   Outside offices rent for
     $150 per month on 1st floor,
     $160 per month on 2nd floor,
     $170 per month on 3rd floor,
       etc.
   Assume all offices are rented. How much rent is collected

   a. each month? b. each year?

Skyscraper

Mathematics teaching objectives:

. Solve problems involving area and money.

Problem-solving skills pupils might use:

. Visualize an object from its description and drawing.

. Break the problem into parts.

. Make a systematic listing.

Materials needed:

. Calculators

Comments and suggestions:

. Encourage pupils to show enough of their work so as to make clear their method of solution.

. Before using the calculator pupils should first set up the problems and indicate the operations to be performed.

. Provide class time for pupils to discuss the procedures they used for getting their answers.

Answers:

1. $2(10 \times 12) + 2(8 \times 12) = 432$ windows

$$432 \times \$88.44 = \$38,206.08$$

2. $36 (2 + 2.5 + 3 + 3.5 + 4 + 4.5 + 5 + 5.5 + 6 + 6.5 + 7 + 7.5) = \$2052.00$

or

$36(2) + 36(2.5) + 36(3) + 36(3.5) + 36(4) + 36(4.5) + 36(5) + 36(5.5) +$
$$36(6) + 36(6.5) + 36(7) + 36(7.5) = \$2052.00$$

(The first method is easier - an application of the distributive property.)

3. a. Inside offices - $125 \times 28 \times 12$

Outside offices - $32(150 + 160 + 170 + 180 + 190 + 200 +$
$$210 + 220 + 230 + 240 + 250 + 260)$$

Total cost - $120,720.00

b. $1,448,640

Grade 7

VIII. PROBABILITY

VIII. PROBABILITY

Probability provides an excellent opportunity
to emphasize these problem-solving skills:
collect data, make a table, use a model, make predictions and in general, guess, check, and refine.
As pupils suggest ways of solving the problems
(or as they describe how they solved the problems)
their answers can be translated into standard
problem-solving phrases. For example, "Keep

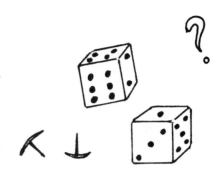

track of what you get" can be verbalized by the teacher as "Make a table."
Specific suggestions for emphasizing the problem-solving skills are given in the
comments and suggestions for each page.

Working with experimental probability in grade 7 is also instructionally
sound. Usually pupils are not ready for any formal probability like "The
probability of rolling a sum of 7 is $\frac{7}{36}$," until somewhere after the seventh
grade. Experimental probability provides some background for later work with
formal probability and, more importantly, it provides experience with a method
of solving problems that is very useful in the real world. To avoid precon-
ceived notions, like "Heads are luckier than tails," many of the activities in
this section use situations with which pupils are not familiar. Each activity
allows pupils to collect enough data to predict the outcome.

Using The Activities

With enough equipment, the entire class can complete one activity and the
results can be compared and combined. Another option is to have small groups
rotate through activity stations which are all on experimental probability.

Materials needed include a pair of regular dice for every two pupils, three
coins for each pupil, a classroom set of scissors, and some mending or scotch tape.

Pupils may need review on an efficient way to tally groups by fives.

It is best if the lessons are presented in the same sequence as given in
the unit. You may wish to provide additional activities in experimental probability. (See Mathematics Resource Project, STATISTICS AND INFORMATION
GATHERING, Creative Publications.)

DIVISION WITH DICE

Rules: Each of 2 players needs a die. Let one player be Player A and the other be Player B.

Each player rolls the die.

Divide A's number by B's number.

   . Player A wins if the first digit of the quotient is 1, 2, or 3.
   . Player B wins if the first digit is 4, 5, 6, 7, 8, or 9.

> ### Make A Guess
>
> Do you think this game is "fair" for Player A? Does Player A have the same chance of winning as Player B?

Play the game 36 times. Record the results in a table showing four things: A's number, B's number, the first digit of the quotient, and the winning player.

FINAL QUESTIONS: Is the game "fair" for Player A?

What digits should Player A have in order to make the game nearly "fair" for both players?

EXTENSION

Play the same game with octahedral dice which have the numbers 1 through 8.

Division With Dice

Mathematics teaching objectives:

. Find quotients involving decimals.
. Determine which of two possibilities is "most likely."
. Record the results of an experiment.

Problem-solving skills pupils _might_ use:

. Identify trends suggested by data in table.
. Make decisions based upon data.
. Create a new problem by varying an old one.

Materials needed:

. A pair of dice for every two pupils
. Octahedral dice (for the extension). See the activity called "Rolling Some Sums."
. Calculators (optional)

Comments and suggestions:

. This game could be used as a total class activity with each pupil working with a partner, or it could be one of several activities used during a lab period.
. Make certain pupils know how to divide, say, 1 by 6 to get a decimal. An alternative is to use calculators to do the division.
. To help determine whether or not the game is fair, pupils should be encouraged to compare their results with others. A teacher-directed discussion of the division possibilities would give pupils another opportunity to see the importance of a systematic listing. Such a listing is given in the table below.

Player A's Roll

	1	2	3	4	5	6
1	①	②	③	4	5	6
2	5	①	①	②	②	③
3	③	6	①	①	①	②
4	②	5	7	①	①	①
5	②	4	6	8	①	①
6	①	③	5	6	8	①

Player B's Roll

The table to the left is a division table showing only the first digit of the quotient. The circled answers indicate the number of opportunities Player A has for winning. If all 36 possibilities were to occur, Player A would win 23 times.

Final question: If the rules were changed so that A would win if the first digit of the quotient were 1 or 2, Player B would have a better chance. Nevertheless, if all 36 possibilities were to occur, Player A still would win 19 times.

Extension: If the same game were played with the octahedral dice, the chances would still favor Player A.

The probability is $\frac{40}{64}$ for Player A

and

$\frac{24}{64}$ for Player B.

ROLLING SOME SUMS

1. Use the pattern below to make an 8-sided die.

2. Work with a partner. Roll both dice; add the two numbers together. What sum did you get?

3. Suppose you rolled the 2 dice many times.
   a. Do you think the sum you rolled in exercise 2 will occur most often?
   b. If not, what sum will occur most often?

4. Roll the dice 64 times. Tally each sum in a table.
   a. Which sum(s) occurred most often?
   b. Which sum(s) occurred least often?
   c. Explain why your results turned out this way.

5. Draw a graph of your results.

EXTENSION

Try to load your dice by taping a small weight to one face. Experiment to see if you were successful.

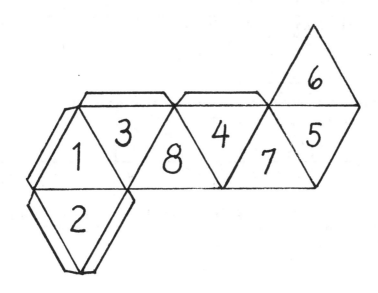

Rolling Some Sums

Mathematics teaching objectives:

. Make a table and graph.
. Determine which possibilities are most likely.
. Make an octahedron.

Problem-solving skills pupils might use:

. Make and use a physical model.
. Identify trends suggested by table or graph.
. Share data and results with other persons.
. Make explanations based upon conditions of the problem.

Materials needed:

. Tagboard
. Scissors
. Scotch or masking tape
. Squared paper for the graph

Comments and suggestions:

. Pupils may need some help in making their eight-sided (octahedral) dice. If several pupils were to be given prior instructions, they could assist the other class members.

. As pupils compare their tables and graphs with others, they should notice certain similarities and differences. If a composite of several tables were made, the likely results would show 9 to be the sum which occurs most often and 2 and/or 16 the sum which occurs least often.

Answers:

3. Check to see that pupils do suggest a sum they think will occur most often.

4. Answers will vary. One actual try is shown:

Sum	2	3	4	5	6	7	8	9	10	11	12	13	14	15	16
No.of times	2	4	4	1	2	8	10	10	6	3	4	6	3	1	0

a. The sums 8 and 9 occurred most often.

b. The sum 16 occurred least often.

c. The table below shows the theoretical number of ways the sums would occur for all 64 possibilities.

Sum	2	3	4	5	6	7	8	9	10	11	12	13	14	15	16
No.of times	1	2	3	4	5	6	7	8	7	6	5	4	3	2	1

5. A bar graph could be made easily on centimetre-squared paper. A collection of pupils' bar graphs on a bulletin board should show the most popular sums to be 8, 9, and 10.

FLIPPING COINS

1. Delores was flipping a coin. She predicted a head or a tail should occur about the same number of times.

   a. Do you agree? _____

   b. Check Delores' prediction by flipping a coin 32 times. Record in the table.

	Tally	Total
Head		
Tail		

2. Mark, Mike, and Tammy were arguing about flipping two coins at the same time.
   - Mark said two heads would occur more often.
   - Mike said two tails would occur more often.
   - Tammy said that one of each should occur just as often as each of the others.

   a. With whom do you agree? _____

   b. Flip two coins 32 times to check the predictions. Record in the table.

	Tally	Total
Two heads		
Two tails		
One of each		

   c. Compare your results with those of your classmates. Explain why the results turn out this way.

3. Arthur was wondering about flipping three coins. He thought two heads and one tail would occur more often.

   a. What do you think the results should be? _____

   b. Flip three coins 32 times to check your prediction.

	Tally	Total
Three heads		
Two heads & one tail		
One head & two tails		
Three tails		

Flipping Coins

Mathematics teaching objectives:

. Determine which possibilities are most likely.
. Record the results of an experiment.

Problem-solving skills pupils _might_ use:

. Guess and check.
. Identify trends suggested by tables.
. Share data and results with other persons.
. Make explanations based upon conditions of the problems.

Materials needed:

. 3 coins for each pupil or for every two pupils

Comments and suggestions:

. Pupils probably will predict correctly for problem 1. However, in problem 2 many pupils will predict Tammy has the correct response only to have the experiment show otherwise.

. Small groups of pupils could make a composite table of the data they collected. The composite tables should show that

- heads occur about as often as tails when a single coin is tossed.

- one head and one tail occurs most frequently when two coins are tossed at the same time.

- two heads and one tail (or two tails and one head) occurs more often when three coins are tossed at the same time.

Answers:

1., 2. and 3. Data in tables will vary. One actual experiment of 32 tosses for each problem is shown. The theoretical probabilities are given in parentheses.

1.

Heads	17 (1 to 2)
Tails	15 (1 to 2)

2.

2 Heads	8 (1 to 4)
2 Tails	10 (1 to 4)
1 of each	14 (1 to 2)

3.

3 Heads	3 (1 to 8)
2 Heads, 1 Tail	12 (3 to 8)
1 Tail, 2 Heads	11 (3 to 8)
3 Tails	6 (1 to 8)

The table below shows the four possible outcomes when 2 coins are flipped:

1st Coin	2nd Coin
H	H
H	T
T	H
T	T

QUIZ OR NO QUIZ

Mr. Snee Key gave these directions to his math class.

. A marker is placed in the top cup. (See diagram on next page.)
. Each pupil needs to flip a coin four times.
. Heads means move down left; tails means move down right.
. If the marker finishes in cups 1, 2, or 5 - NO QUIZ.
. If the marker finishes in cups 3 or 4 - QUIZ.

Did Mr. Key make up a fair problem?   Use a marker.
Try it 16 times.   Record the results in the table.

Trials	Flips				Finishing Cup	Quiz? YES or NO
	1	2	3	4		
1						
2						
3						
4						
5						
6						
7						
8						
9						
10						
11						
12						
13						
14						
15						
16						

1. How many times did you have NO QUIZ? _____   QUIZ? _____
2. How many times did the class have NO QUIZ? _____   QUIZ? _____
3. Do you think Mr. Key made a fair deal? _____
4. Explain why the results turn out this way.
5. Take the four flips for Trial 1 and rearrange the order of the flips. Now where does the marker finish?

Quiz Or No Quiz

Mathematics teaching objectives:

. Determine which possibilities are most likely.

. Record the results of an experiment.

Problem-solving skills pupils might use:

. Identify trends suggested by tables.

. Share data and results with others.

. Make explanations based upon the conditions of the problem.

Materials needed:

. Markers and coins

Comments and suggestions:

. You may need to clarify the diagram shown on the second page of the activity.

. Let pupils work in pairs with each helping the other read and interpret directions, collect, and record data.

. After partners compare their results with others, a total class tally should be made of the number of trials which resulted in a quiz or no quiz.

. The **actual** probability for a quiz is 5 out of every 8 trials.

Answers:

One table filled with experimental data is shown.

1. No quiz - 7;

    a quiz - 9.

2. Answers will vary.

3. No

4. To land in either cup 1 or cup 5, a pupil must flip

Trials	Flips				Finishing Cup	Quiz? Yes or No
	1	2	3	4		
1	H	T	H	H	2	No
2	H	T	T	H	3	Yes
3	T	H	T	T	4	Yes
4	T	T	T	T	5	No
5	H	H	H	H	1	No
6	T	T	T	H	4	Yes
7	H	T	H	H	2	No
8	H	T	H	T	3	Yes
9	H	H	H	T	2	No
10	T	H	T	H	3	Yes
11	H	H	T	T	3	Yes
12	H	H	T	T	3	Yes
13	H	H	H	T	2	No
14	T	H	H	T	3	Yes
15	H	H	T	T	3	Yes
16	H	H	H	T	2	No

4 heads in a row or 4 tails in a row. This is not likely to happen.

5. The rearrangement of the four plays in trial 1 all result in landing in the same finishing cup. This will be true for any of the trials.

Possible Extension: Change the rules so that the results of the problem will make the possibilities of a quiz or no-quiz equally likely.

Answers: If the marker finishes in cups 1, 3, and 5 - no quiz.
If the marker finishes in cups 2 and 4 - quiz.

QUIZ OR NO QUIZ

Heads

Tails

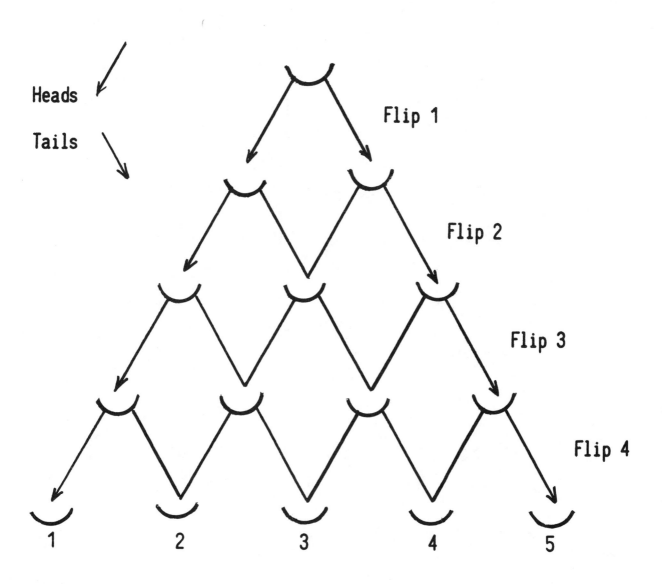

Flip 1

Flip 2

Flip 3

Flip 4

1   2   3   4   5

THE COIN FLIP PATH

1. If you flip a coin 50 times, about how many heads and how many tails would you expect to get? Heads ____   Tails ____

2. Now flip 50 times. Record H or T below. How many heads and tails did you actually get?   Heads ____   Tails ____

___ ___ ___ ___ ___ ___ ___ ___ ___ ___

___ ___ ___ ___ ___ ___ ___ ___ ___ ___

___ ___ ___ ___ ___ ___ ___ ___ ___ ___

___ ___ ___ ___ ___ ___ ___ ___ ___ ___

___ ___ ___ ___ ___ ___ ___ ___ ___ ___

3. Use the 50 coin flips to make a coin flip path. Tails means move ↘ . Heads means move ↙ . Not counting the start, how many times would you expect to return to the center line? ____

Start here

4. Trace the path indicated by your 50 coin flips.

5. How many times did you return to the center line? __

6. If the coin flips alternated H, T, H, T, H,..., how many times would you return to the center line?__

7. Look at your coin flips. Circle the series of flips that caused you to move farthest away from the center line.

8. Trace the path indicated by the coin flips but in reverse order; that is, from flip 50 down to flip 1. Will you return to the center line the same number of times?

Idea from Statistics and Information Organization. Creative Publications
© PSM 81

The Coin Flip Path

Mathematics teaching objectives:

. Be aware of variations possible in a sequence of two equally likely events.

. Record the results of an experiment.

Problem-solving skills pupils might use:

. Record solution attempts.

. Identify trends suggested by a graph.

. Make a general statement based on the results of an investigation.

Materials needed:

. A coin for each pupil

Comments and suggestions:

. This activity should be easier after working the Quiz Or No Quiz activity.

. Let each pupil work the activity individually prior to a discussion in which the paths are compared.

. The main value of the activity is for pupils to compare the coin-flip paths. Pupils should find that the path returns to the center line much less than expected.

Answers:

1. Answers will vary, but 25 heads and 25 tails would be the best answer.

2. The record of flips will vary. One sequence with 19 heads and 31 tails is given below:

```
T  T  T  H  H  T  H  H  T  H
T  H  T  H  T  T  T  H  H  T
T  T  T  T  H  H  T  H  T  T
T  T  T  T  T  H  T  H  T  T
T  H  T  H  H  H  T  T  H  T
```

3. Many pupils will think 25 times is a good guess. Actually, this number of returns is highly unlikely. The coin-flip path for the above sequence moves a long way from the center line.

5. 4 times for the above sequence.

6. 25 times. (Any divergence from this alternating pattern will reduce the number of times the path returns to the center line.)

7. In the above sequence, the largest number of successive tails (7) results in the longest move away from the center line. (Notice this tail sequence starts on the tail side of the center line.)

8. The above sequence does, but others may not return the same number of times.

Grade 7

IX. CHALLENGES

IX. CHALLENGES

The activities in the Getting Started section were very directed and pupils were encouraged to use (although not completely restricted to) one problem-solving skill at a time. The challenge problems in this section leave the choice of the problem-solving method up to the pupil. The intention is to allow for and encourage individual differences, creativity, and cooperation.

Let's look at how one challenge problem, "Money In The Bank" (page 233), can be used in the classroom. As you read the example, notice how the teacher does not structure or direct the methods pupils use, but that the teacher does have these important functions:

. to help pupils understand the problem.

. to listen if pupils want to discuss their strategies.

. to praise and encourage pupils in their attempts, successful or not.

. to facilitate discussion of the problem and sharing of the strategies.

. to give hints or ask questions, if necessary.

. to summarize or emphasize methods of solution after pupils have solved the problem.

As we look into Ms. Walters classroom, the page, "Money In The Bank," has been distributed. Pupils are used to seeing a different challenge each week and they know they will be given some time during the week to work on the problem.

Ms. W: Here is the challenge problem for the week. I'll let you look at the problem, then we'll discuss it to be sure we all understand it. (Waits) Wayne, you look puzzled.

Wayne: It seems to me that after just one month Arthur will have three times as much as Jane.

Ms. W: But don't forget that each of them starts with $10 - not $0.

Kay: Ms. Walters, I don't really know how to get started.

Ms. W: Look at the list of problem-solving skills posted on the board. Which one might you use?

Kay: I think I'll try guessing.

Ms. W: Good idea. (In a quiet, side conversation with Jeff) Jeff, you evidently found that 5 months was too small. How did you come to that conclusion?

Jeff: Well, I made a guess of 5 and found that Jane would have $15 and Arthur would have $25. Twenty-five dollars is not twice as much as fifteen - but I'm getting closer.

Ms. W: That's good! Keep trying. (Ms. Walters continues to walk around the room. More individual discussions and work takes place. Several describe what they did and Ms. Walters summarizes as necessary.) I've noticed that several of you are using a systematic list and looking for patterns. The rest of you may wish to try this approach when you work on the problem later in the week. Be sure to let me know how you're progressing.

The above approach to challenge problems also gives opportunities for practicing the following problem-solving skills:

. State the problem in your own words.

. Clarify the problem through careful reading and by asking questions.

. Share data and the results with other interested persons.

. Listen to persons who have relevant knowledge and experiences to share.

. Study the solution process.

. Invent new problems by varying an old one.

THE TEACHER MUST BE AN ACTIVE, ENTHUSIASTIC SUPPORTER OF PROBLEM SOLVING.

<u>Using The Activities</u>

Sixteen varied challenge problems are provided. Some teachers give them as a "challenge of the week" or as a Friday activity. On the day the challenge is given out, time should be spent on getting acquainted with the problem. On following days, a few minutes can be devoted to pupil progress reports. If there is little sign of progress, you can provide some direction by asking a key question or suggesting a different strategy. At appropriate times, the activity can be summarized by a class discussion of strategies used and some problem extensions.

<u>One</u> <u>Plan</u> <u>For</u> <u>Using</u> <u>A</u> <u>Challenge</u> (over a period of 1 or 2 weeks)

First day -

- . Give out the challenge. (Possibly near the end of the period.)

- . Let pupils read written directions and possibly discuss with
  a classmate.

- . Clarify any vocabulary which seems to be causing difficulty. Ask
  a few probing questions to see if they have enough understanding
  to get started.

- . Remind them that during a later math class, time will be used to
  look at the problem again.

Later in the week -

- . Have pupils share their ideas.

- . Identify the problem-solving skills suggested by these ideas.

- . Conduct a brainstorming session if pupils do not seem to know
  how to get started.

- . Suggest alternative strategies they might try.

- . Give an extension to those pupils who have completed the challenge.

On a subsequent day -

- . Allow some class time for individuals (or small groups) to work
  on the challenge. Observe and encourage pupils in their attempts.

- . Try a strategy along with the pupils (if pupils seem to have given up.)

Last day -

- . Conduct a session where pupils can present the unsuccessful as well
  as the successful strategies they used.

- . Possibly practice a problem-solving skill that is giving pupils difficulty; e.g., recording attempts, making a systematic list, or checking
  solutions.

Key problem-solving strategies that pupils have used in solving the problems are given in the comments for each problem. Your pupils might have
additional ways of solving the problems.

A challenge problem for the teacher: Keep the quick problem solver from
telling answers to classmates.

MONEY IN THE BANK

Jane and Arthur each have
$10 in the bank.

Every month Jane plans to
add $1 to her account.

Arthur plans to add $3 to
his account every month.

1. In how many months will
   Arthur have twice as
   much as Jane?

2. Now solve the problem if they each start with $15 rather
   than $10.

3. Solve the problem if they each start with $20.

4. Write about anything you've discovered.

EXTENSION

If they each start with $10, when will Arthur have
three times as much as Jane?

Money In The Bank

Problem-solving skills pupils might use:

. Make a systematic list.

. Make predictions based upon observed patterns.

. Guess and check.

Comments and suggestions:

. To get them started, some pupils may need to see the beginnings of
  an organized list:

	Jane	Arthur
At start	$10	$10
After 1 month	$11	$13
After 2 months	$12	$16

. Some pupils will probably use a guess, check, and refine method. This
  method should be discussed when summarizing the solution processes.
  This allows an excellent opportunity to remind pupils of this powerful
  problem-solving skill.

Answers:

1. 10 months
2. 15 months
3. 20 months
4. Answers will vary, but the most common statement likely will be:
       The number of months is equal to the number of dollars
       in the bank at the beginning.

Extension: Not possible.
            Pupils often try the extension by making the same type of listing
            as before. After much effort, they realize that as they continue
            they are getting nearer to a solution. However, the solution is
            always "out of their grasp." You might guide the discussion by
            asking about what happens after 100 months, 1000 months, 1,000,000
            months, 1,000,000,000 months, etc. These examples show how the
            amount is getting closer and closer to three times as much but
            never will get there.

Months	Jane	Arthur
100	110	310
1,000	1,010	3,010
1,000,000	1,000,010	3,000,010
1,000,000,000	1,000,000,010	3,000,000,010

COUNTY FAIR

During a recent county fair, Bull's-Eye Bill was paid 20¢ each time he hit the target at the rifle range. He paid 50¢ each time he missed.

1. How many hits did Bill have if he lost $4.70 in 15 shots?

2. In 15 shots how many hits must Bull's-Eye have to win more money than he loses?

3. In 15 shots, can he win exactly $4.20?

4. In 15 shots, can he lose exactly $4.20?

EXTENSION

Suppose Bull's-Eye shot 30 times. How many hits must he have to win more money than he loses?

County Fair

Problem-solving skills pupils <u>might</u> use:

 . Guess and check.

 . Make a systematic list.

 . Look for patterns.

Comments and suggestions:

 . You may need to clarify the problems by giving an example showing a possible result when 15 shots are taken.

 . The problems can be solved by using a guess and check procedure. However, in summarizing the solutions it would be good to show a systematic listing of all possibilities. (See below) The patterns in the listing should also be discussed.

Answers:

1. 4

2. At least 11

3. No

4. No

Extension:

 . At least 22

Shots Made	Money Made	Shots Missed	Money Lost	Net
0	$ 0	15	$7.50	-7.50
1	$ .20	14	$7.00	-6.80
2	$ .40	13	$6.50	-6.10
3	$ .60	12	$6.00	-5.40
4	$ .80	11	$5.50	-4.70
5	$1.00	10	$5.00	-4.00
6	$1.20	9	$4.50	-3.30
7	$1.40	8	$4.00	-2.60
8	$1.60	7	$3.50	-1.90
9	$1.80	6	$3.00	-1.20
10	$2.00	5	$2.50	- .50
11	$2.20	4	$2.00	+ .20
12	$2.40	3	$1.50	+ .90
13	$2.60	2	$1.00	+1.60
14	$2.80	1	$ .50	+2.30
15	$3.00	0	$ 0	+3.00

WHO'S WHO?

1. Four married couples belong to a golf club. The wives' names are Kay, Sally, Joan, and Ann. Their husbands are Don, Bill, Gene, and Fred.

   Examine the following clues. They should help you decide who is married to whom?

   . Bill is Joan's brother.

   . Joan and Fred were once engaged, but "broke up" when Fred met his present wife.

   . Ann has two brothers, but her husband is an only child.

   . Kay is married to Gene.

   A chart like the one below can help you sort out the clues.

	Kay	Sally	Joan	Ann
Don				
Bill				
Gene				
Fred				

2. Steve, Jim, and Calvin are married to Beth, Donna, and Jane, not necessarily in that order. Four of them are playing bridge.

   . Jim does not play bridge.
   . Donna's husband and Steve's wife are partners.
   . Jane's husband and Beth are partners.
   . No married couples are partners.

   a. Who is married to whom?
   b. Who are partners?

Who's Who?

Problem-solving skills pupils might use:

. Use a chart to keep track of information.

. Eliminate possibilities.

Comments and suggestions:

. In solving logic problems, one needs to find a system for keeping track of and sorting out information. A chart is often the best way to accomplish this. Pupils may need to be shown how such a chart can be used.

Answers:

1.

	Kay	Sally	Joan	Ann
Don	X	X	Yes	X
Bill	X	Yes	X	X
Gene	Yes	X	X	X
Fred	X	X	X	Yes

2. a.

	Beth	Donna	Jane
Steve	X	X	Yes
Jim	Yes	X	X
Calvin	X	Yes	X

Jane's husband and Donna's husband play bridge. Therefore, Beth is married to Jim who doesn't play bridge.

Jane is married to Steve (since Steve's wife and Donna's husband are bridge partners and no married couples are partners).

Donna is married to Calvin (the only possibility left).

b. Jane and Calvin are bridge partners. Steve and Beth are bridge partners.

More logic problems can be found in The Oregon Mathematics Teacher - November 1977 and February 1978.

-238-

JARS OF CANDY

These five jars of candy contain a total of 100 pieces.

How many pieces of candy are
in each jar if

1. each jar contains two more pieces
   than the previous jar? (Place
   your answers at the right.)

   __ __ __ __ __

2. each jar contains four more
   pieces than the previous one?

   __ __ __ __ __

3. each jar contains six more
   than the previous one?

   __ __ __ __ __

4. each jar contains eight more
   pieces than the previous one?

   __ __ __ __ __

5. each jar contains ten more
   pieces than the previous one?

   __ __ __ __ __

EXTENSION

What if we wanted the 100 pieces in six jars rather than five?
Investigate to see when this is possible.
Use the same clues as before.

<u>Jars</u> <u>Of</u> <u>Candy</u>

Problem-solving skills pupils <u>might</u> use:

. Guess and check.

. Look for patterns.

Comments and suggestions:

. Pupils may need to be reminded of two conditions -

- that the total number of pieces is 100 and

- each jar contains a different amount of candy.

. In discussing the solutions, you might emphasize how the concept
of an average can be helpful in obtaining the answers.

Answers:

1.	16	18	20	22	24	
2.	12	16	20	24	28	
3.	8	14	20	26	32	
4.	4	12	20	28	36	
5.	0	10	20	30	40	Note: Pupils may balk at having a jar with 0 pieces.

Extension:

No solution is possible, unless fractional pieces of candy are allowed.
Some pupils may be challenged to determine what fractions will work.

CHANGE FOR A DOLLAR

Sammy works at the amusement park. Usually he is making change all evening. But one night business was slow. To amuse himself, Sammy figured out different ways to make change for one dollar.

Sammy found that it was easy to make change for a dollar using 2 coins, 3 coins, 4 coins, and 5 coins.

Number of Coins	50¢	25¢	10¢	5¢	1¢
2	✓✓				
3	✓	✓✓			
4		✓✓✓✓			
5	✓	✓	✓✓	✓	
6					
⋮					

See if you can extend Sammy's chart all the way down to 20 coins. Can you extend it any further? Try it.

Change For A Dollar

Problem-solving skills pupils might use:

. Guess and check.

. Make a systematic list.

. Identify patterns suggested by data in a table.

Comments and suggestions:

. There are many possible ways to complete the table. One way which shows some interesting patterns is given below.

. The table can be filled in consecutively up to 77 coins. You may wish to have a large chart on the bulletin board on which pupils can place their solutions.

No. of Coins	50¢	25¢	10¢	5¢	1¢
2	2	-	-	-	-
3	1	2	-	-	-
4	-	4	-	-	-
5	1	1	2	1	-
6	1	1	1	3	-
7	1	1	-	5	-
8	-	3	-	5	-
9	-	2	3	4	-
10	-	-	10	-	-
11	-	-	9	2	-
12	-	-	8	4	-
13	-	-	7	6	-
14	-	-	6	8	-
15	-	-	5	10	-
16	-	-	4	12	-
17	-	-	3	14	-
18	-	-	2	16	-
19	-	-	1	18	-
20	-	-	-	20	-

Note the exchanges that have been made and the patterns that show up in the lower part of the table.

STEPPING HIGH

1. Calvin is building a staircase pattern as
    shown in the figure. Each block is one
    foot high. How many blocks would it
    take to build steps that would be
    20 feet high?

    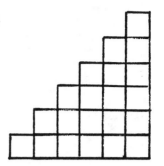

    a. Describe two or more methods Calvin could use to solve his
    problem.

    b. Use one of your methods to solve the staircase problem.
    Check your solution by using a second method.

    c. Suppose Calvin's staircase is to be 22 feet high. Now
    how many blocks are needed?

    d. Suppose Calvin has 131 blocks to use. What would be the
    height of a staircase he could build?

2. Calvin's brother has longer legs.
    His staircase looks like this.
    How many blocks are needed for a
    staircase that is 20 feet high?

    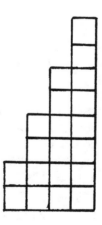

Problem-solving skills pupils <u>might</u> use:

- Use a drawing.
- Use a systematic list (table).
- Make generalizations based upon observed patterns.
- Solve easier but related problems.

Comments and suggestions:

- You might provide cubes, tiles, or graph paper for those pupils who need a more visual approach.

- Pupils will no doubt use strategies that combine several of the skills listed above. In summarizing the solutions you should make certain all of the strategies pupils used are discussed. An important strategy that needs special attention is the first one described below.

Answers:

1. a. Some strategies pupils might use:

    - Use simpler cases.
        Organize the information in a table.
        Look for patterns.
        Extend the pattern to the 20th case. (Pupils will probably discover different ways to extend the table. These should be discussed.)

Height	1	2	3	4	5	6	7	
Number of Blocks	1	3	6	10	15			

    - Recognize that the problem can be solved by finding the sum of the numbers from 1 to 20.

    - Draw a picture and then count the number of blocks needed.

    b. 210 blocks   c. 253 blocks   d. 15 feet   2. 110 blocks

Possible Extension: ⌐ How many blocks are needed if Calvin's staircase ⌐
                         is to have a height of 100 feet?   ⌐

- This problem can be solved by any of the methods described above. However, each of these methods would be tedious. A general rule would be much more efficient.
  By examining the above table, you'll note that $N = \dfrac{H(H+1)}{2}$
  Notice that this formula does <u>not</u> rely on any previous part of the table.

- A drawing method can also be used to arrive at this formula. Note that the rectangle in Figure B has been formed by duplicating the figure in A. This rectangle has dimensions of N and (N + 1). It's area must be divided by 2 since the number of blocks has been doubled.

Fig. A

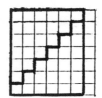

Fig. B

PIGS AND TURKEYS

Mr. Farmer raises pigs and turkeys.

One day he noticed that –

      the total number of heads and wings was equal
      to the total number of feet.

How many pigs and how many turkeys does Mr. Farmer have?

Find as many solutions to this problem as you can.

<u>Pigs</u> <u>and</u> <u>Turkeys</u>

Problem-solving skills pupils <u>might</u> use:

. Guess and check.

. Make a systematic list.

. Make a drawing.

Comments and suggestions:

. Pupils may guess at the number of pigs and turkeys; check their guess and refine accordingly.

. Other pupils may make a systematic list showing various possibilities and continue until a correct solution is found. (see below)

Answer:

The smallest combination is 1 pig and 3 turkeys.

PIGS		TURKEYS				
Heads	Feet	Heads	Feet	Wings	Heads + Wings	Feet
1	4	1	2	2	4	6
1	4	2	4	4	7	8
1	4	3	6	6	10	10
2	8	1	2	2	5	10
2	8	2	4	4	8	12
2	8	3	6	6	11	14
2	8	4	8	8	14	16
2	8	5	10	10	17	18
2	8	6	12	12	20	20

There are an infinite number of solutions. If P is the number of pigs, then 3P is the number of turkeys $(P \neq 0)$.

WHAT'S THE PATH?

Tammie made up this challenge problem.

> Find a path so the sum of the numbers along the path is 121.
>
> Your path must go through the open gates and must end in the circle.
>
> The path cannot go through a box more than once.

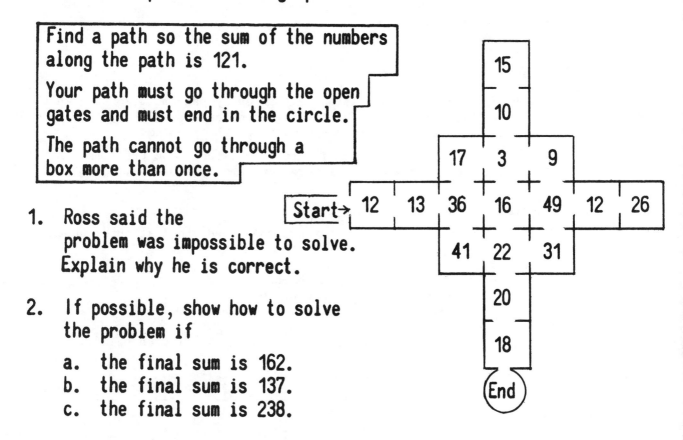

1. Ross said the problem was impossible to solve. Explain why he is correct.

2. If possible, show how to solve the problem if

   a. the final sum is 162.
   b. the final sum is 137.
   c. the final sum is 238.

3. There are a total of 8 different paths that will give 8 different sums. See if you can find all 8 of them. Record the numbers used and the totals.

EXTENSION

Use the figure at the right. Create your own challenge problem. Make it so there are three different paths that will end up with 41. All the numbers must be different.

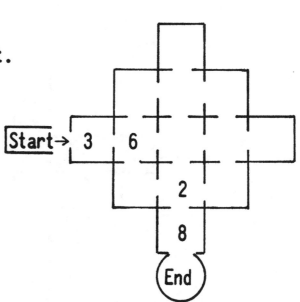

<u>What's The Path</u>?

Problem-solving skills pupils <u>might</u> use:

- Guess and check.
- Work backwards.
- Make and use a systematic list.

Comments and suggestions:

- A very useful observation is that every path must go through the 12, 13, and 36 boxes at the beginning and the 22, 20, and 18 boxes at the end. This greatly simplifies the problem.
- The extension allows pupils to create their own problems. A bulletin board display would be interesting and informative.

Answers:

1. The sum of the first three numbers and the last three numbers is already 121.

2. a) See the first column below.
    b) See the second column below.
    c) Impossible. The greatest possible sum is 237.

3. The 8 different paths are systematically listed below:

12	12	12	12	12	12	12	12
13	13	13	13	13	13	13	13
36	36	36	36	36	36	36	36
41	16	16	16	17	17	17	17
22	22	3	49	3	3	3	3
20	20	9	31	16	16	9	9
18	18	49	22	22	49	49	49
162	137	31	20	20	31	16	31
		22	18	18	22	22	22
		20	217	157	20	20	20
		18			18	18	18
		229			237	215	230

Extension: One possible solution -

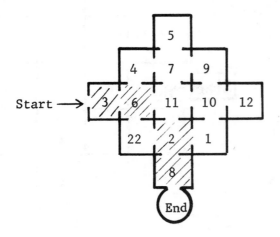

The three paths are:

3	3	3
6	6	6
22	4	11
2	7	10
8	11	1
41	2	2
	8	8
	41	41

SPIROLATERALS - 1

Spirolaterals are designs made by a sequence of numbers. The sequence 1,2,3 is called an order-3 spirolateral because it has 3 terms.

1. To draw a 1,2,3 spirolateral

    a. mark a starting point.

    b. move <u>1</u> space <u>up.</u>

    c. move 2 spaces <u>right.</u>

    d. move <u>3</u> spaces <u>down.</u>

    e. move <u>1</u> space <u>left.</u>

    f. move <u>2</u> spaces <u>up.</u>

    g. move <u>3</u> spaces <u>right.</u>

    Continue until you get back to the starting point.

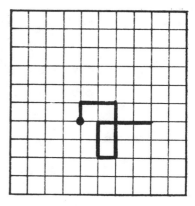

2. Draw a 4,2,1 and a 2,4,3, spirolateral. Remember, the key to drawing any spirolateral is UP-RIGHT-DOWN-LEFT.

> The rectangles in the 1,2,3 meet at a point in the middle. Call this <u>Type A</u>.
>
> The rectangles in the 4,2,1 have an open space in the middle. Call this <u>Type B</u>.
>
> The rectangles in the 2,4,3 overlap in the middle. Call this <u>Type C</u>.

3. Draw four order-3 spirolaterals of your own. Record the sequence of numbers and classify each as Type A, B, or C.

4. Describe a number pattern that identifies a

    (a) Type A spirolateral   (b) Type B spirolateral

    (c) Type C spirolateral

Spriolaterals - 1

Problem-solving skills pupils <u>might</u> use:

. Make and use a drawing.

. Make generalizations based upon observed patterns.

Comments and suggestions:

. Each pupil needs several sheets of graph paper.

. Pupils will need to be reminded of the UP - RIGHT - DOWN - LEFT
movement. Perhaps these "directions" could be posted on the
chalk board.

Answers:

1, 2, and 3:   Student drawings.

4.   For an order-3 spirolateral, if the sum of the two smallest numbers
is:
  a) equal to the third number   - Type A
  b) less than "   "   "   - Type B
  c) greater than the third number - Type C

Possible Extensions:

. Can you determine the dimensions of the resulting rectangles before
drawing the spirolateral? (The rectangles will have dimensions equal
to the two smallest numbers in the sequence.)

. Can you determine the number of squares in the middle of a Type B
spirolateral or the number of squares in the overlap of a Type C
spirolateral? (Subtract the sum of the two smallest numbers from
the third one, and then square this difference. Notice how this
would be zero for a Type A spirolateral indicating no open space or
no overlap.

. What happens when you rearrange the sequence of numbers in an order-3
spirolateral?

A 1,2,3 spirolateral is called an order-3 spirolateral because it has 3 numbers in its sequence.

> How many loops through the numbers does it take to complete this spirolateral?

> Do all spirolaterals take the same number of loops as the order-3 one?

> Do all spirolaterals return to the starting point?

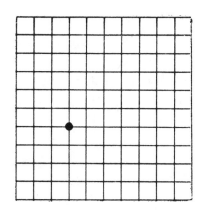

1. Investigate these questions by making the spirolaterals given in the table below. You may use others if you wish.

SPIROLATERAL SEQUENCE	ORDER	NUMBER OF LOOPS
1	1	
1,2	2	
1,2,3	3	
1,2,3,4		
1,2,3,4,5		
1,2,3,4 5,6		
1,2,3,4,5,6,7		
1,2,3,4,5,6,7,8		

2. Describe any patterns you found in the table.

3. How many loops would it take to complete these spirolaterals?

   a. order-11

   b. order-24

   c. order-18

   d. order-72

Problem-solving skills pupils <u>might</u> use:

  . Make and use a drawing.

  . Make generalizations based upon observed patterns.

Comments and suggestions:

  . Each pupil needs several sheets of graph paper.

  . Pupils will need to be reminded of the questions they are attempting
    to answer in the first part of this investigation. Once the table
    is completed and the questions are answered, it is likely that pupils
    will make further conjectures that lead to more investigation.

Answers:

1.

SPIROLATERAL SEQUENCE	ORDER	NUMBER OF LOOPS
1	1	4
1,2	2	2
1,2,3	3	4
1,2,3,4	4	does not return
1,2,3,4,5	5	4
1,2,3,4,5,6	6	2
1,2,3,4,5,6,7	7	4
1,2,3,4,5,6,7,8		does not return

2. One pattern in the number of loops is: 4 - 2 - 4 - does not - 4 - 2 - 4 - does not -

3. a) 4
   b) does not
   c) 2
   d) does not

GENERALIZATIONS:

  . Odd order spirolaterals take 4 loops.
  . Even order (but not divisable by 4) spirolaterals take 2 loops.
  . Even order (and divisable by 4) spirolaterals do not return to the starting point.

Possible Extension:

Pupils enjoy making spirolaterals from their zip codes, telephone numbers,
house numbers, or their names by assigning values according to this table.

1	2	3	4	5	6	7	8	9
A	B	C	D	E	F	G	H	I
J	K	L	M	N	O	P	Q	R
S	T	U	V	W	X	Y	Z	

FOUR 4's

Use four 4's and the operations of $+$, $-$, $\times$, and $\div$.

Write expressions for as many of the numbers from 0 to 20 as you can. Two are already finished for you.

0 =   11 =

1 =   12 =

2 =   13 =

3 =   14 =

4 =   15 =

5 =   16 =

6 =   17 = $(4 \times 4) + (4 \div 4)$

7 =   18 =

8 = 4 + 4 + 4 - 4   19 =

9 =   20 =

10 =

EXTENSION

Find expressions for any of the numbers from 21 to 100.

<u>Four 4's</u>

Problem-solving skills pupils <u>might</u> use:

. Guess and check.

. Use appropriate mathematical notation.

Comments and suggestions:

. Emphasize the importance of using proper mathematical symbolism.

. You may wish to have the pupils make a bulletin board display of different solutions.

Answers:

. Answers will vary.

. A possible solution for each problem is given.

. Note that several solutions involve square roots or decimals. You may want to tell pupils that this is allowable.

$$0 = 4 + 4 - 4 - 4$$

$$1 = (4 + 4) \div (4 + 4)$$

$$2 = \frac{4}{4} + \frac{4}{4}$$

$$3 = (4 + 4 + 4) \div 4$$

$$4 = \frac{4 - 4}{4} + 4$$

$$5 = (4 \times 4 + 4) \div 4$$

$$6 = \frac{4 + 4}{4} + 4$$

$$7 = 4 + 4 - \frac{4}{4}$$

$$8 = 4 + 4 + 4 - 4$$

$$9 = 4 + 4 + \frac{4}{4}$$

$$10 = \frac{4}{.4} + 4 - 4$$

$$11 = \frac{4}{.4} + \frac{4}{4}$$

$$12 = 4 \times 4 - \sqrt{4} - \sqrt{4}$$

$$13 = \frac{44}{4} + \sqrt{4}$$

$$14 = 4 + 4 + 4 + \sqrt{4}$$

$$15 = 4 \times 4 - \frac{4}{4}$$

$$16 = \frac{4 \times 4 \times 4}{4}$$

$$17 = (4 \times 4) + (4 \div 4)$$

$$18 = \frac{4}{.4} + 4 + 4$$

$$19 = \frac{4 + 4 - .4}{.4}$$

$$20 = (4 + \frac{4}{4}) \times 4$$

SOME CIRCLE PUZZLES

1. Miss Young has her 18 pupils seated in a circle. They are evenly spaced and numbered in order.

    a. Which pupil is directly opposite pupil number 1?
    b. Which pupil is directly opposite pupil number 5?
    c. Which pupil is directly opposite pupil number 18?

2. Mr. Evans seated his pupils in the same way as Miss Young's.
   Pupil number 5 is directly opposite number 16.
   How many pupils are in Mr. Evan's class?

3. Mrs. White teaches Phys. Ed. She had her pupils space themselves evenly around a circle and then count off.
   Pupil number 16 is directly opposite number 47.
   How many pupils are in Mrs. White's class?

4. A huge number of boys are standing in a circle and are evenly spaced. The 7th boy is directly opposite the 791st.
   How many boys are there altogether?

Some Circle Puzzles

Problem-solving skills pupils <u>might</u> use:

. Make and use a drawing.

. Look for patterns.

Comments and suggestions:

. You might need to suggest to some pupils that they arrange slips of paper around a circle.

. Some discussion of the meaning of "directly opposite" may be necessary.

. You might have pupils use the face of a clock to help them discover patterns suggested by "directly opposite" numbers.

. Remind pupils that if they can discover some general rules the problems can be solved much more efficiently. This is expecially true for problem 4 due to the size of the numbers.

Answers:

1. a. 10   b. 14   c. 9

   These answers are easy for pupils to determine after they discover that:

   . if $\underline{n}$ is the number of the person with the highest number, the directly-opposite person has a number of $\frac{n}{2}$.

   . the difference between the numbers of directly-opposite persons is always the same.

2. 16 - 5 = 11;   11 x 2 = <u>22</u>, the number of pupils in Mr. Evan's class.

3. 47 - 16 = 31;   31 x 2 = <u>62</u>, the number of pupils in Mrs. White's class.

4. 791 - 7 = 784;   784 x 2 = 1568 boys

POOL PATTERNS

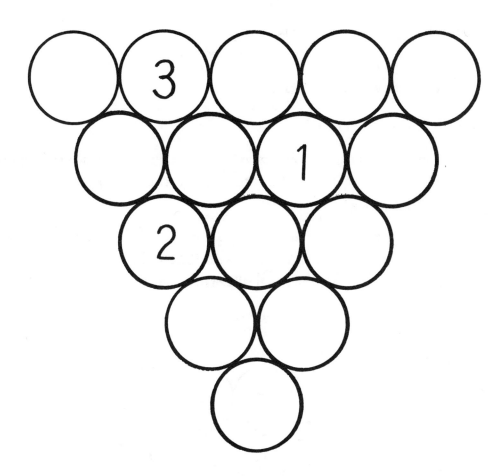

Oregon Skinny discovered some interesting ways to arrange pool balls. In one arrangement, each number can be determined by subtracting the two numbers immediately above.

Suppose Skinny tells you the location of the 1, 2, and 3 balls (see diagram above). Determine the location of the other balls.

EXTENSION

Skinny tried another arrangement. This time he tried to make each number equal to the sum of the two immediately above. How could you convince Skinny that this arrangement is impossible. (The 1, 2, and 3 balls do not have to be in the same location as in the diagram above.)

Pool Patterns

Problem-solving skills pupils <u>might</u> use:

. Guess and check.

. Eliminate possibilities.

Comments and suggestions:

. You might suggest that pupils make 12 small "markers" numbering
  them from 4 to 15. This will facilitate the guess, check, and
  refine procedure.

. One way to solve the problem is to begin with the top, left circle
  and keep trying different possibilities. (This is often referred
  to as solving a problem by "exhaustion.")

Answer:

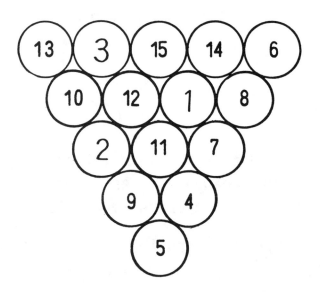

Extension: After some trial and error, one finds that

- the 15-ball has to be at the bottom and 14 and 1
  have to be directly above.

- But the 1-ball has to be in the top row.

- Impossible. The 1-ball cannot be in both locations.

THE BOY AND THE DEVIL

There once was a boy who spent all his time dreaming about getting rich. "I would do any-thing to be rich," he said. The devil, hearing this, appeared before the boy and offered to make him rich.

"See that bridge?" said the devil. Just walk across and I will double the money you have now. In fact, each time you cross I will double your money. There is just one thing--you must give me $24 after each crossing.

The boy agreed. He crossed the bridge, stopped to count his money and, sure enough, it had doubled. He paid the devil $24 and crossed again. Again his money doubled. He paid another $24 and crossed a third time. Again his money had doubled, but this time there was only $24 left which he had to pay the devil so he was left with nothing. The devil laughed and vanished.

1. How much money did the boy start with?

2. Suppose the problem remains the same except that the boy gives up $30 rather than $24. Now how much did the boy start with?

3. Suppose the problem remains the same except that the boy crosses four times rather than three times. (He still gives up $24 each time.) Now how much did the boy start with?

The Boy And The Devil

Problem-solving skills pupils might use:

- Guess and check.
- Keep an organized record of trials.
- Work backwards.

Comments and suggestions:

- To get some pupils started you may need to suggest a guess, check, refine approach. By "running through" a particular guess they should begin to understand the structure of the problem and also decide whether their guess was too large or too small.

- A good way to summarize the guess, check, refine process is to use a chart.

Before Crossing		After Crossing
1st trip	30	2(30) - 24 = 36
2nd trip	36	2(36) - 24 = 48
3rd trip	48	2(48) - 24 = 72

This shows that a guess of $30 is too large.

Before Crossing		After Crossing
1st trip	20	2(20) - 24 = 16
2nd trip	16	2(16) - 24 = 8
3rd trip	8	2(8) - 24 = -8

This shows that a guess of $20 is too small.

- The working backward skill can also be summarized by using the same kind of chart and filling in the blanks starting at the end.

Before Crossing		After Crossing		
1st trip	☐	☐	- 24 =	☐
2nd trip	☐	☐	- 24 =	☐
3rd trip	☐	24	- 24 =	0

Answers:

1. $21   2. $26.25   3. $22.50

CONNECTED STAMPS

You have 8 stamps connected together like this:

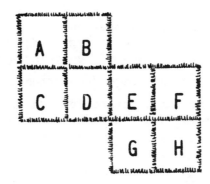

1. List all the ways 2 connected stamps can be torn off.

2. List all the ways 3 connected stamps can be torn off.

3. List all the ways 4 connected stamps can be torn off.

4. List all the ways 5 connected stamps can be torn off.

5. List all the ways 6 connected stamps can be torn off.

6. List all the ways 7 connected stamps can be torn off.

EXTENSION

Draw a pattern of 8 connected stamps. Make it such
that 3 connected stamps can be torn off in 17 ways.

Connected Stamps

Problem-solving skills pupils might use:

. Look for geometric patterns.
. Make a systematic listing.

Comments and suggestions:

. Pupils will probably discover one or the other of the skills listed
above as being the most efficient. You may need to suggest one of
these methods to pupils who are having difficulty.

. This activity provides a good opportunity to discuss geometric symmetry.

. Two different solution strategies are shown below for problems 1 and 2.
You should discuss both methods when summarizing the results.

Answers:

1. 9 ways.

    1st strategy - Look for geometric patterns.

    In the ABCD square, there are 4 different
    "2-stamp" rectangles.
    In the EFGH square, there are 4 different
    "2-stamp" rectangles.
    DE is another 2-stamp rectangle.

    2nd strategy -

    Make a systematic listing.

    AB
    AC
    BD
    CD
    DE
    EF
    EG
    FH
    GH

2. 12 ways.

    1st strategy: In the ABCD square there are 4 different 3-stamp figures.
                    In the CDEF rectangle there are 2 different 3-stamp figures.
                    In the EFGH square there are 4 different 3-stamp figures.
                    BDE and DEG are 2 different 3-stamp figures.

    2nd strategy: ABC   BCD   CDE   DEF   EFG   FGH
                    ABD   BDE   DEG   EFH
                    ACD   EGH

3. 12 ways   (ABCD, ABDE, ACDE, BCDE, BDEF, BDEG, CDEF, CDEG, DEFG, DEGH,
                  DEFH, EFGH)

4. 14 ways   (ABCDE, ABDEF, ABDEG, ACDEF, ACDEG, BCDEF, BCDEG, BDEFG, BDEFH,
                  BDEGH, CDEFG, CDEFH, CDEGH, DEFGH)

5. 13 ways   (ABCDEF, ABCDEG, ABDEFG, ABDEFH, ABDEGH, ACDEFG, ACDEFH, ACDEGH,
                  BCDEFG, BCDEFH, BCDEGH, BDEFGH, CDEFGH)

6. 6 ways   (ABCDEFG, ABCDEFH, ABCDEGH, ABDEFGH, ACDEFGH, BCDEFGH)

Extension:

. There are three 2 by 2 squares.
. Each has four different 3-stamp figures. (12)
. There are four 1 by 3 rectangles. (4)
. Finally, there is EFH. (1)

A SHOPPING PUZZLE

How many people went shopping?

These clues should help you decide –

- . Each person had $1.00 to spend.
- . Each person did not have to spend the whole dollar.
- . Each person bought only pens and pencils.
- . Each person bought at least one of each.
- . Each person bought a different combination of pens and pencils.

How many different combinations can you find?

Rate yourself –

Less than 12	–	??
13–20	–	O.K.
21–26	–	Good
27–29	–	Fantastic
Over 29	–	New Record

EXTENSION

Use the same clues but change the prices.
Create a "pens and pencils" puzzle that has only 5 solutions.

A Shopping Puzzle

Problem-solving skills pupils <u>might</u> use:

- . Guess and check.
- . Make a systematic list.
- . Break the problem into parts.

Comments and suggestions:

- . You may need to clarify the conditions of the problem by giving some examples that work and some that are not allowable.
- . Pupils may start listing possibilities by making guesses, checking, discarding the combinations that do not work, and recording the ones that do. Such a procedure would result in many different workable combinations. However, a more systematic approach would provide more assurance that all possible combinations were identified. One such systematic approach is shown below.

Answer: 29 people went shopping (29 different combinations)

One Pencil		Two Pencils		Three Pencils		Four Pencils		Five Pencils	
No.of Pens	Total Cost	No.of Pens	Total Cost	No.of Pens	Total Cost	No.of Pens	Total Cost	No.of Pens	Total Cost
1	24	1	33	1	42	1	51	1	60
2	39	2	48	2	57	2	66	2	75
3	54	3	63	3	72	3	81	3	90
4	69	4	78	4	87	4	96		
5	84	5	93						
6	99								

Six Pencils		Seven Pencils		Eight Pencils		Nine Pencils	
No.of Pens	Total Cost	No.of Pens	Total Cost	No.of Pens	Total Cost	No.of Pens	Total Cost
1	69	1	78	1	87	1	96
2	84	2	93				
3	99						

Extension:

Answers will vary. One example:

Pens cost 55¢ each and pencils cost 9¢ each. If the clues remain the same as given in the challenge, there are only 5 combinations.